POSTCARD HISTORY SERIES

Beaver Dam

IN VINTAGE POSTCARDS

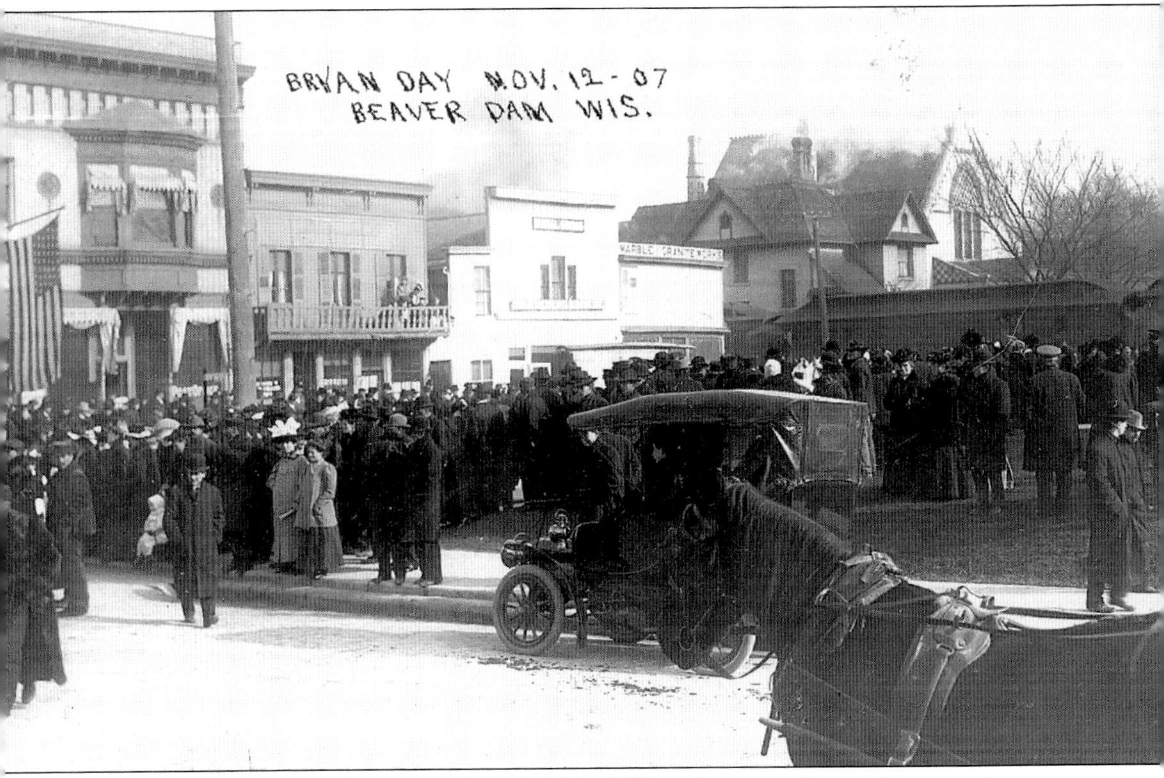

BRYAN DAY. On Tuesday, November 12, 1907, William Jennings Bryan, the Great Commoner, visited the city. All schools and businesses were closed so that people could attend the speech, which lasted 1 hour and 15 minutes. Bryan urged the public not to panic during the current recession, noting that the country was on the rebound. He also reversed his controversial 1896 position of returning to the gold standard.

POSTCARD HISTORY SERIES

Beaver Dam

IN VINTAGE POSTCARDS

Roger Noll

ARCADIA PUBLISHING

ISBN 978-0-7385-3973-7

Published by Arcadia Publishing
Charleston, South Carolina

Printed in the United States of America

Library of Congress Catalog Card Number: 2005928684

For all general information, contact Arcadia Publishing:
Telephone 843-853-2070
Fax 843-853-0044
E-mail sales@arcadiapublishing.com
For customer service and orders:
Toll-free 1-888-313-2665

Visit us on the Internet at www.arcadiapublishing.com

On the cover: **THE SWAN CITY PARK PAVILION.** Built in 1881 as part of the Vita Park Spa, the Swan City Park Pavilion has been an enduring Beaver Dam attraction since its inception. The pavilion has gone through many changes over the years, but in 1985, the landmark was restored to its original splendor and placed on the National Register of Historic Places.

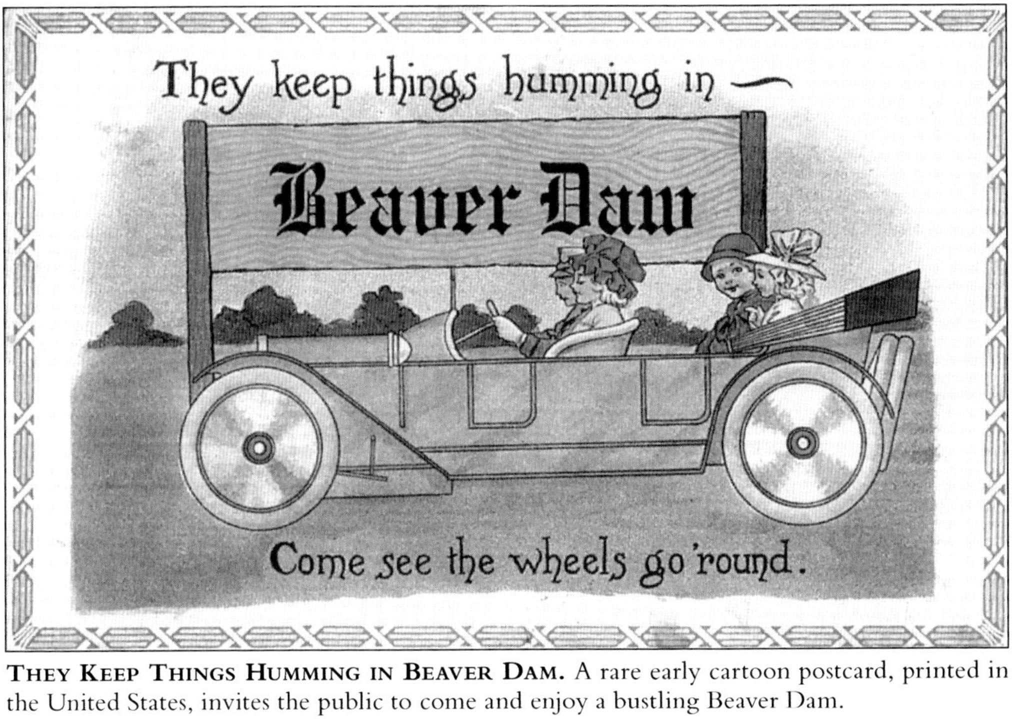

THEY KEEP THINGS HUMMING IN BEAVER DAM. A rare early cartoon postcard, printed in the United States, invites the public to come and enjoy a bustling Beaver Dam.

Contents

Acknowledgments

The creation of any book requires the assistance of many supporters. This is especially applicable to a tome dedicated to the American postcard. The collecting of postcards has been a growing hobby since the golden era of the picture postcard at the beginning of the 20th century. In creating a collection significant to a particular area, one is bound to find a core assortment of postcards that is common to all collectors. Yet each collector will, through luck or persistent searching, find those few dozen rare postcards that are coveted by all collectors. Therefore, in order to create a truly unique publication of postcards, it is necessary to rely on the cooperation of other ardent collectors. In my case, I was fortunate to have been allowed access to four other collections. I sincerely thank Roger Pearson, Myron Schrauth, Kathleen Binder, and Scott Frinak for entrusting me with the Beaver Dam postcards that have taken them a lifetime to accumulate.

But the postcards are of little benefit without the historical explanations provided by the captions. My research was greatly assisted by the fine staff of the Beaver Dam Community Library. I extend a special thank-you to local historian and librarian Pat Pawl.

Finally, I would like to thank my mother and father, Henry and Bernadette Noll. Their unwavering support has always been an inspiration to attain whatever success I have achieved through the years.

Introduction

In 2003, in cooperation with the Dodge County Historical Society, I authored *Beaver Dam: 1841–1941*, published as part of Arcadia Publishing's Images of America series. That pictorial history of Beaver Dam's first century was written to celebrate the 65th anniversary of the historical society.

The present volume seeks to honor another significant occasion in the city's history, the sesquicentennial observance of Beaver Dam's charter incorporation as a city in 1856. A companion piece to the earlier book, which focused on the people and history of the city, this edition uses vintage postcards to represent the landmarks of Beaver Dam.

The American postcard began in 1893, when the federal post office issued cards of the Columbian Exposition in Chicago. This gave way to private mailing cards in 1898. The era from the turn of the century to the start of World War I ushered in the golden era of postcards. With the introduction of the common real-picture postcard, the use of postcards exploded. Post office records show the number of postcards mailed at the end of fiscal year (June 30, 1908) as being 677,777,798. This is staggering considering the population at the time was merely 88,700,000.

The real-picture postcard gave the viewer a snapshot of local landmarks and events, and the postcard was the e-mail of its day. For the mere cost of the card and 1¢ postage, the sender could advise the reader of an upcoming visit, an update on a relative's health, or confirmation that a gift was received. Amazingly, the tourist postcard with the written message "Having a great time. Wish you were here" was rarely seen in the golden era.

Yet, as important as the message was to the writer and recipient at the time, it is the postcard face (whether a photograph or cartoon) that fascinates the historian now. At the dawn of the 20th century, cameras were the domain of the professional photographer. It was he who sought out the grand buildings, street scenes, storefronts, parks, beachfronts, and novel activities to be placed on card fronts. The historian holds a great debt to the professional postcard photographer whose work preserved the classic images of small-town America. And it is ironic that a medium designed to be used and immediately discarded has become one of the major archives of our early daily life.

I hope you enjoy this look at Beaver Dam, Wisconsin, as memorialized in vintage postcards.

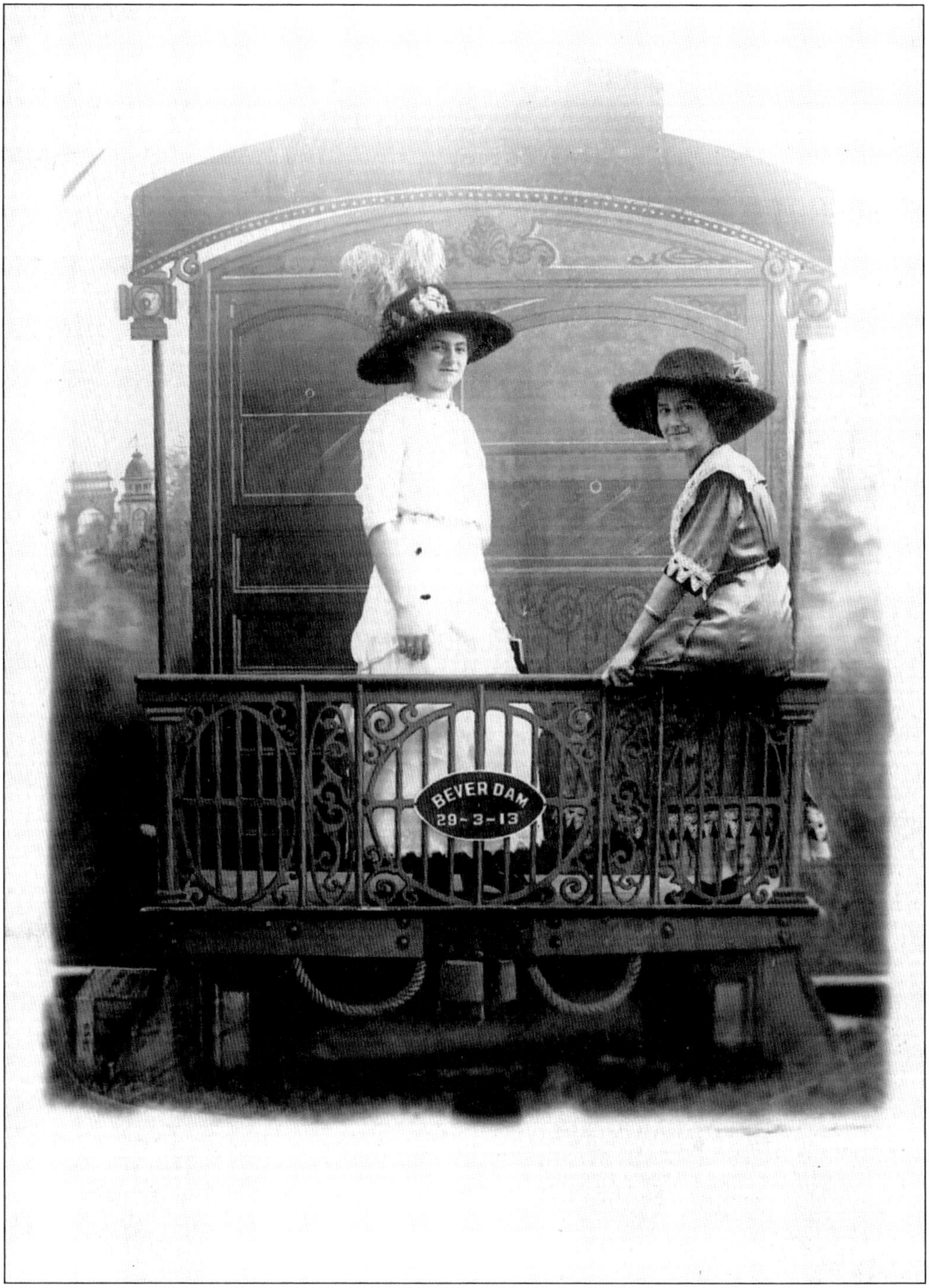

A Studio Memory. This type of postcard was popular just prior to World War I. A backdrop was placed in a photographer's studio in such a manner as to allow the subject to give the impression of being inside a moving vehicle. Common drops included biplanes, zeppelins, trains, and automobiles. Although the studio is not identified, only three existed in Beaver Dam at the time—L. Hartford, Veling and Brothers, and William Huebner.

One

Get Off the Earth and Come to Beaver Dam

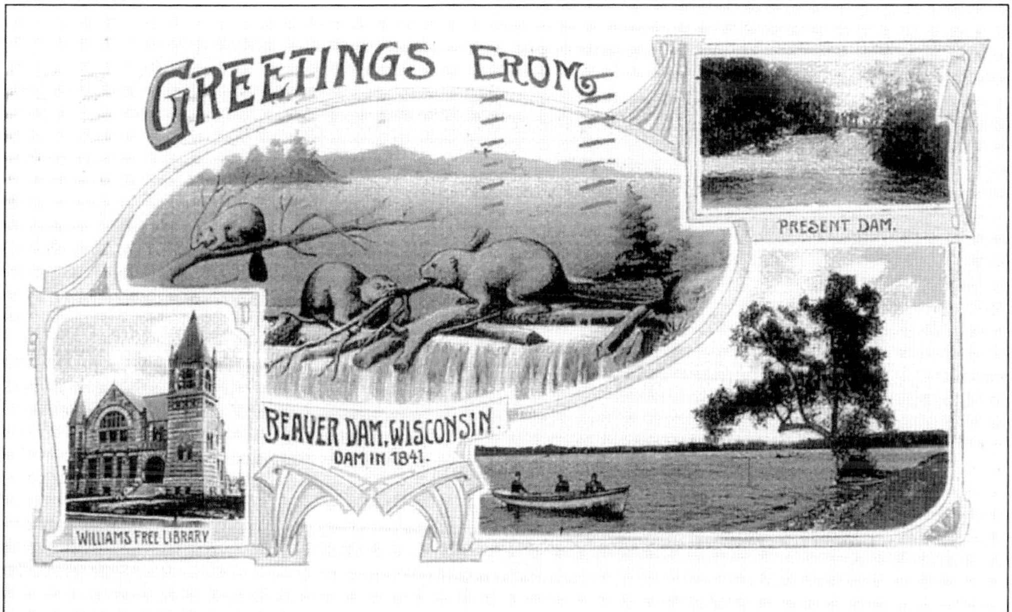

Greetings from Beaver Dam. This multicolored postcard, postmarked in 1910, depicts several images of interest in the city: the dam, the lake, the Williams Free Library, and a scene of a beaver building a dam. The beaver scene was used to symbolize the city's beginning throughout Beaver Dam's first 100 years.

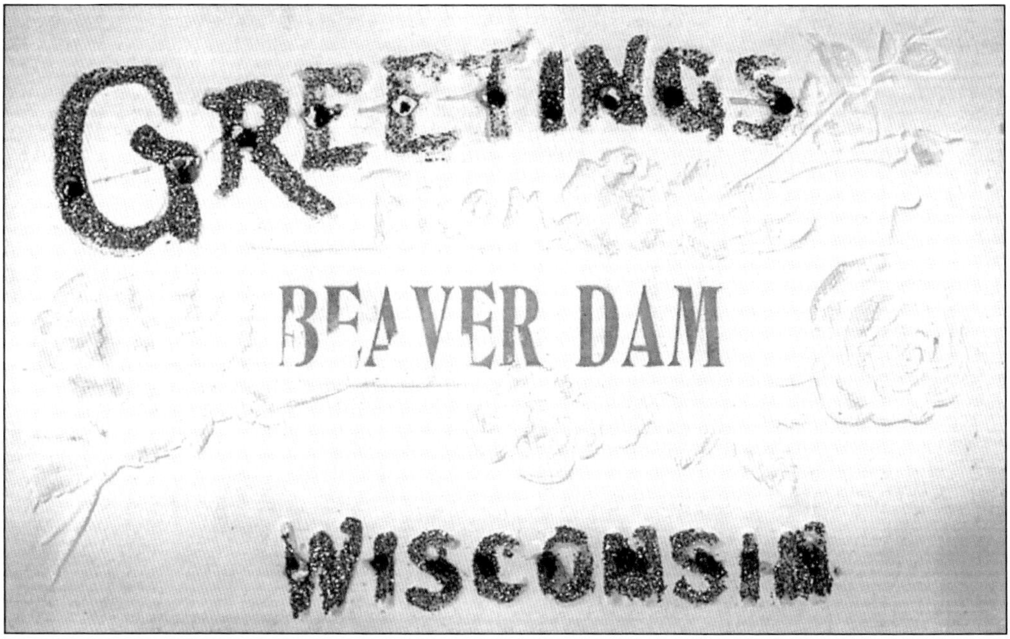

THE GERMAN INFLUENCE. The Germans dominated the postcard industry at the beginning of the 20th century. Their prowess in lithographic printing allowed for economic production of four-color postcards, whether real-photo postcards or these three-dimensional postcards.

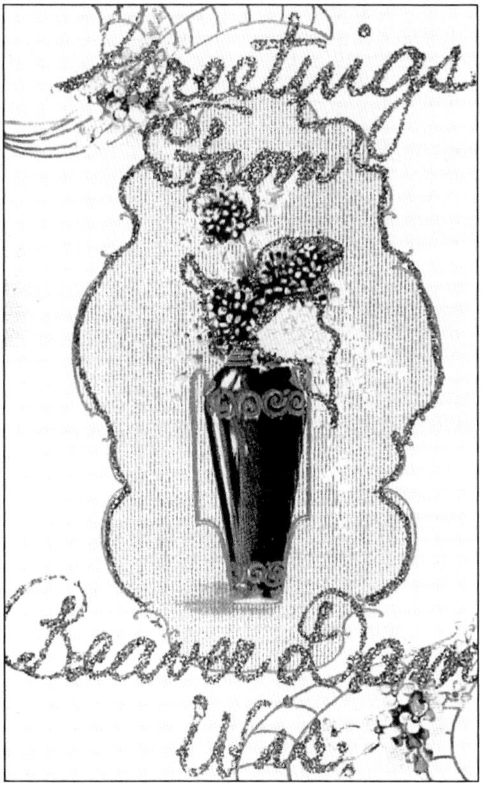

THE RAISED IMPRINT. The raised impressions of the German postcards on this page are accented by the bright colors and by glitter attached to the lettering.

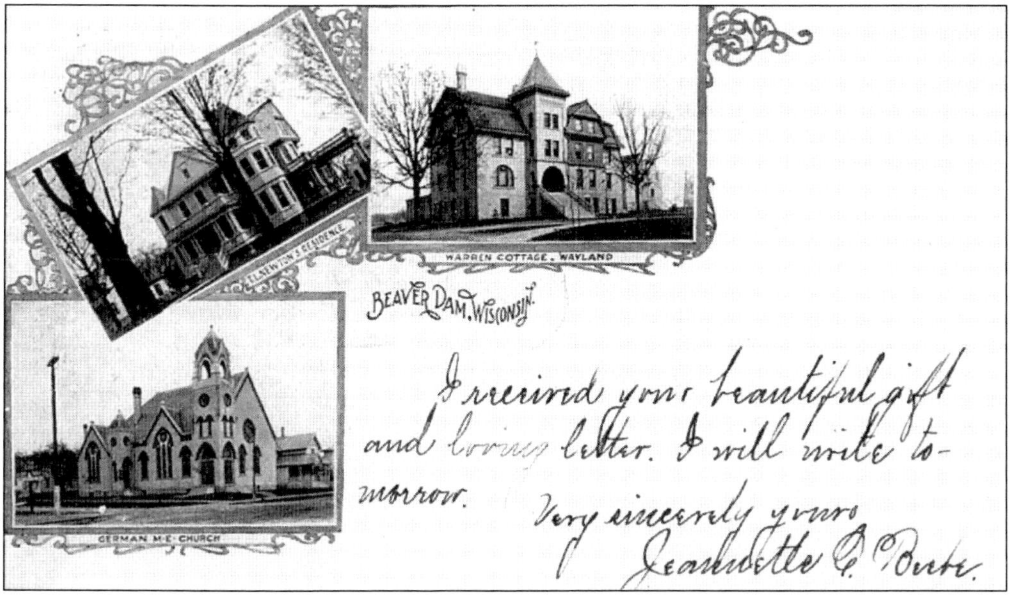

A Pre-1907 Postcard. Prior to 1907, writing was not allowed on the back of postcards. To address this restriction, many postcards provided space on the front for a short note. The sender of this postcard, Jeannette C. Beebe, was an instructor at Wayland Academy.

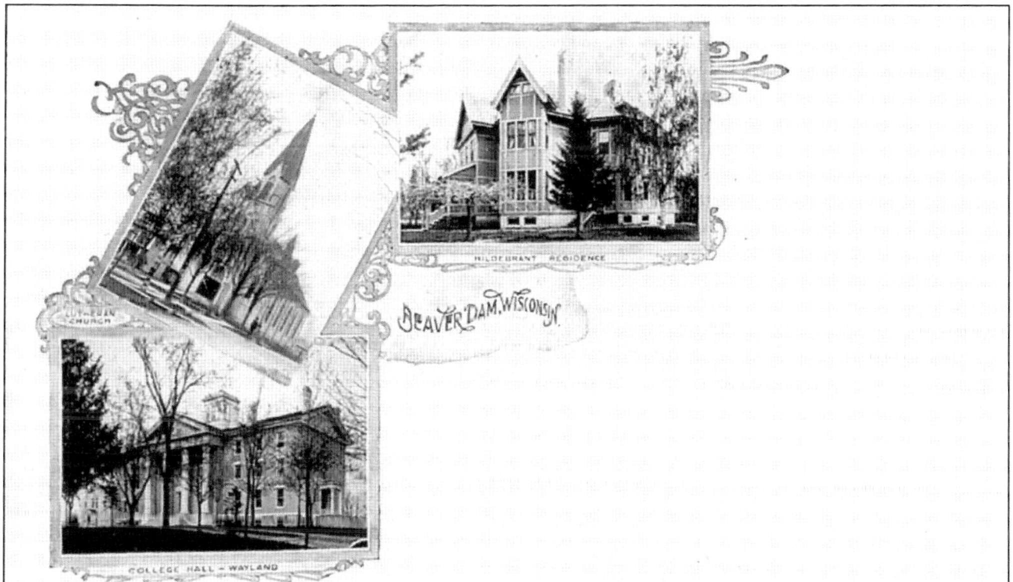

The Multi-image Postcard. With the advancement of black-and-white photographic printing, postcard publishers experimented with multi-image postcards. The photographs displayed various landmarks of interest in any given city.

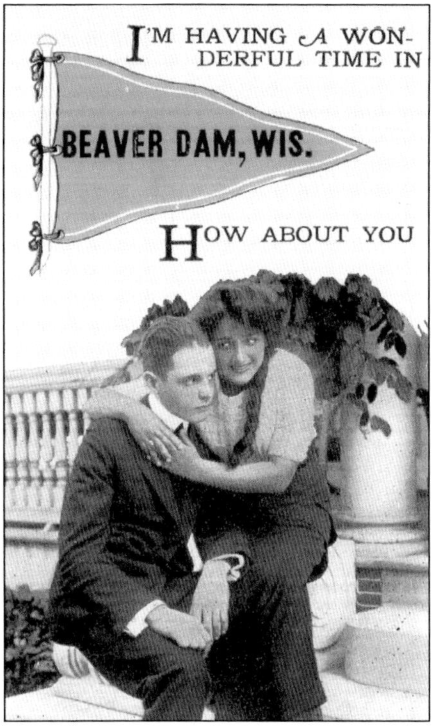

THE PENNANT POSTCARD. The booster postcard, or pennant postcard, came into popularity just prior to World War I. It was intended to proclaim the many benefits and activities available throughout the city. However, the photographs used were generic in nature.

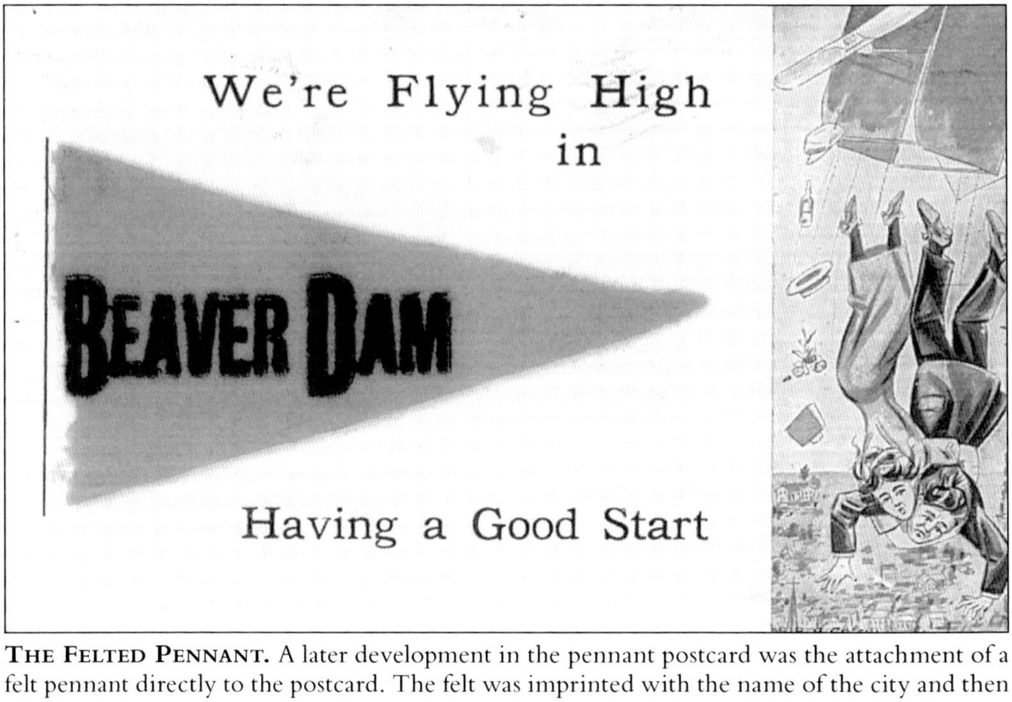

THE FELTED PENNANT. A later development in the pennant postcard was the attachment of a felt pennant directly to the postcard. The felt was imprinted with the name of the city and then glued on by hand. The felt pennant postcard was phased out over the years because it was costly, labor intensive, and heavy, and it could not be sorted by hand or machine.

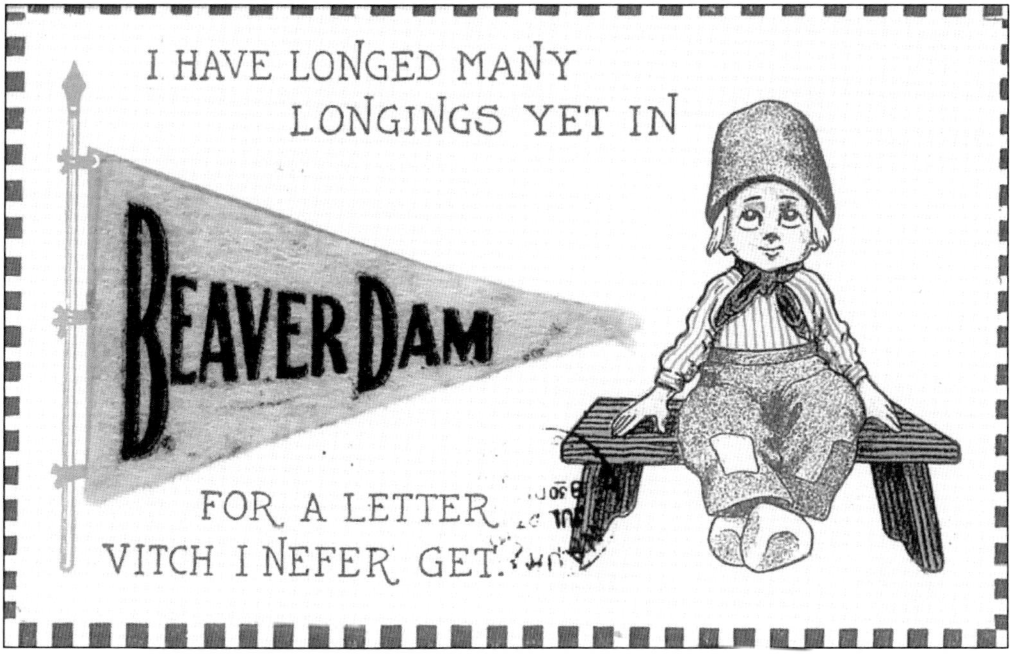

The Influence of Immigration. These 1913 postcards recognize the great European immigration to the nation's Midwest. The two postcards pictured here, with their broken English phraseology, depict the Dutch influence.

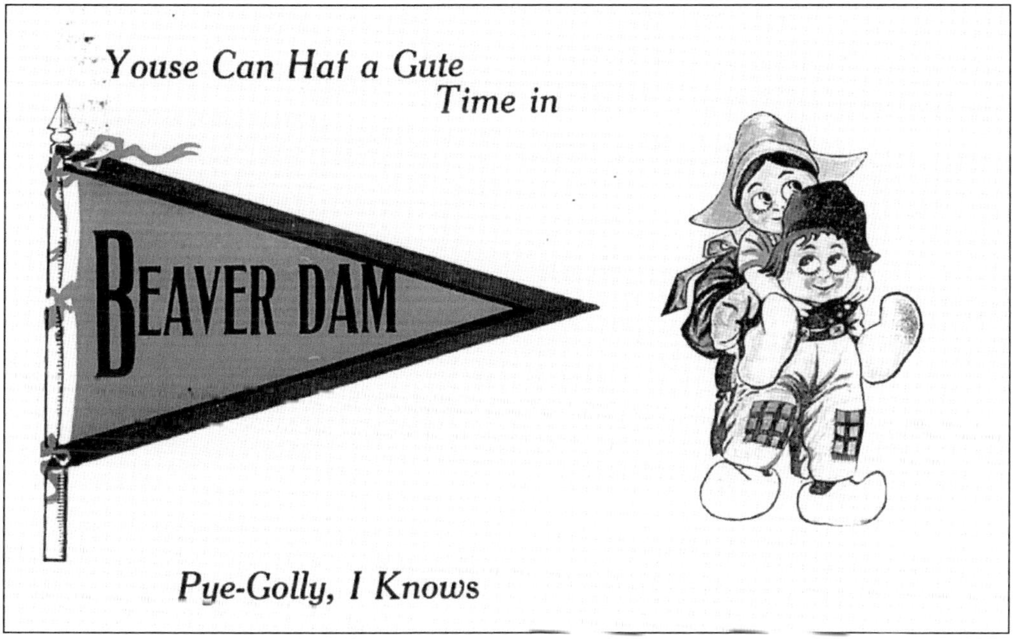

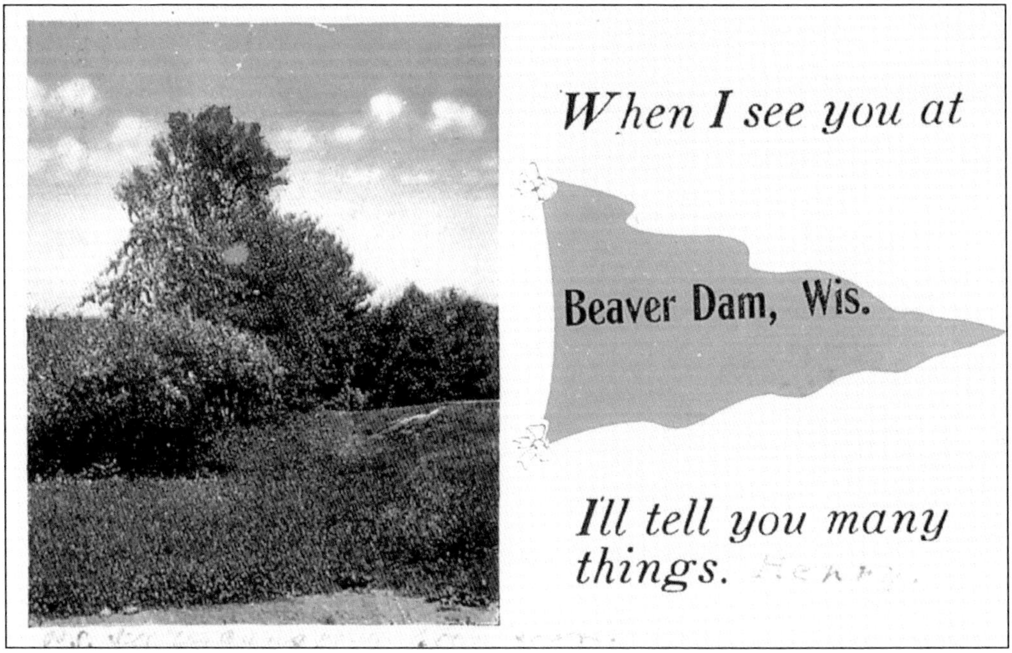

I'll Tell You Many Things. This 1918 postcard combines the landscape and pennant styles of postcards.

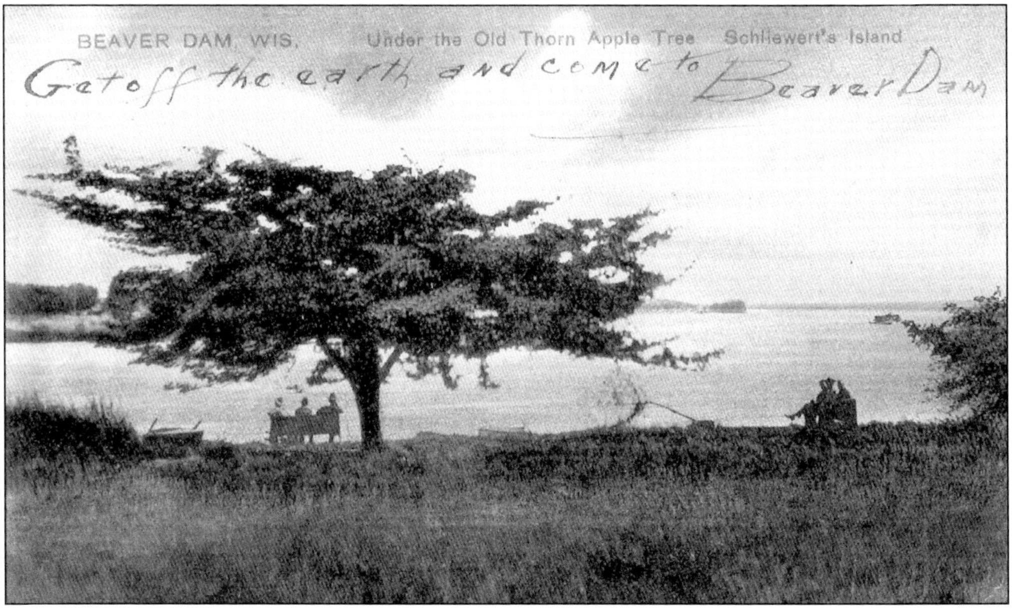

Get Off the Earth and Come to Beaver Dam. This is an excellent example of the German lithographic landscape postcard. The special hand-printed message to William Johnson of Stevens Point lends itself to two interpretations. Either Johnson should visit Beaver Dam (a place so rare that it cannot be found on this planet), or he needs to get up off his backside and journey to Beaver Dam.

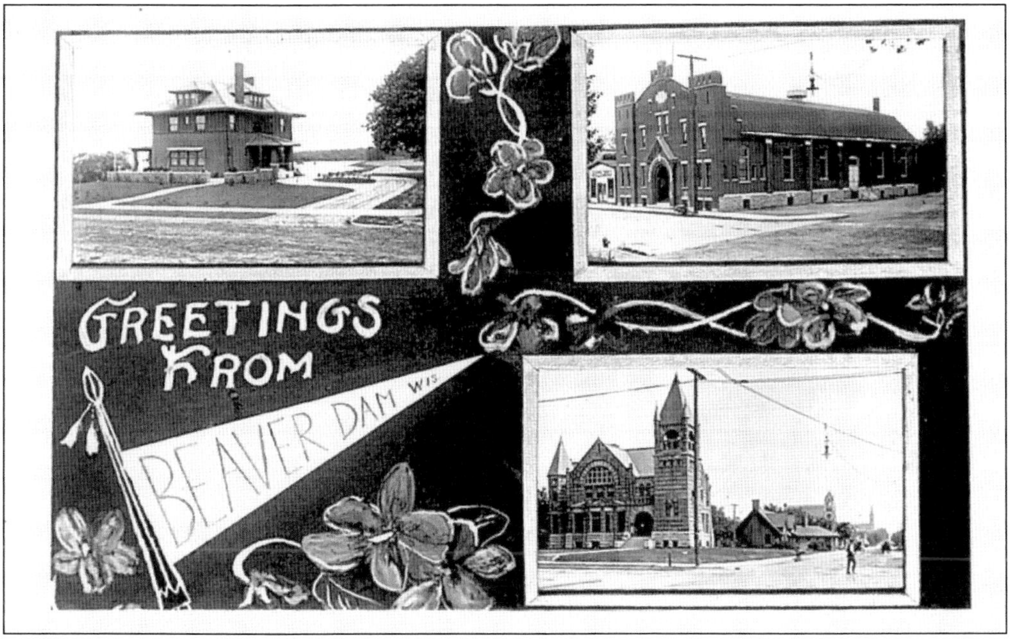

GREETINGS FROM THE 1920S. The multi-photo postcard saw a resurgence in the 1920s, messages were now allowed on the backs of postcards, leaving the entire face free for photographs. This postcard publisher produced a high-quality postcard with sophisticated graphics.

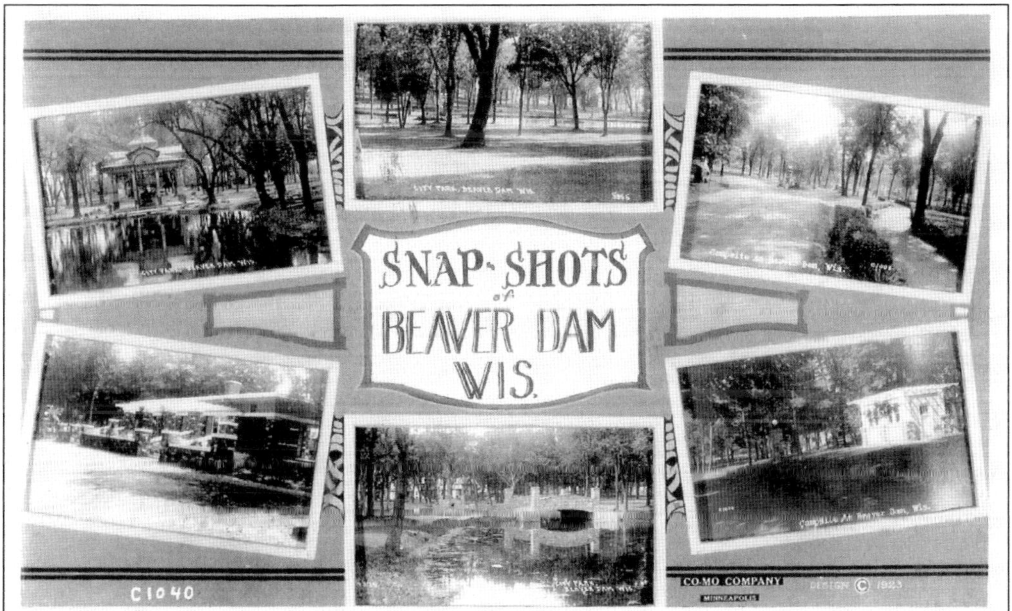

CITY PARK GREETINGS. Locations like City Park, which offered a number of subjects of photographic interest, were well suited to the multi-photo postcard of the 1920s.

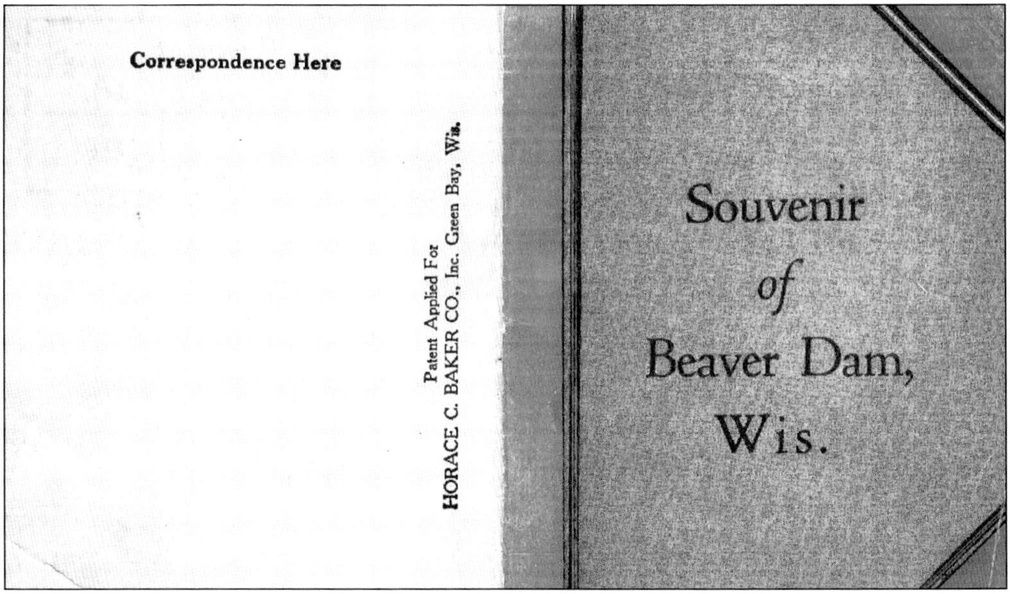

A Novelty Postcard of the 1930s. So novel was the design of this souvenir postcard that its creator, the Horace C. Baker Company, applied for a patent. The postcard's front was comprised of a small booklet that the reader opened to reveal a multiple-page travelogue of Beaver Dam sites.

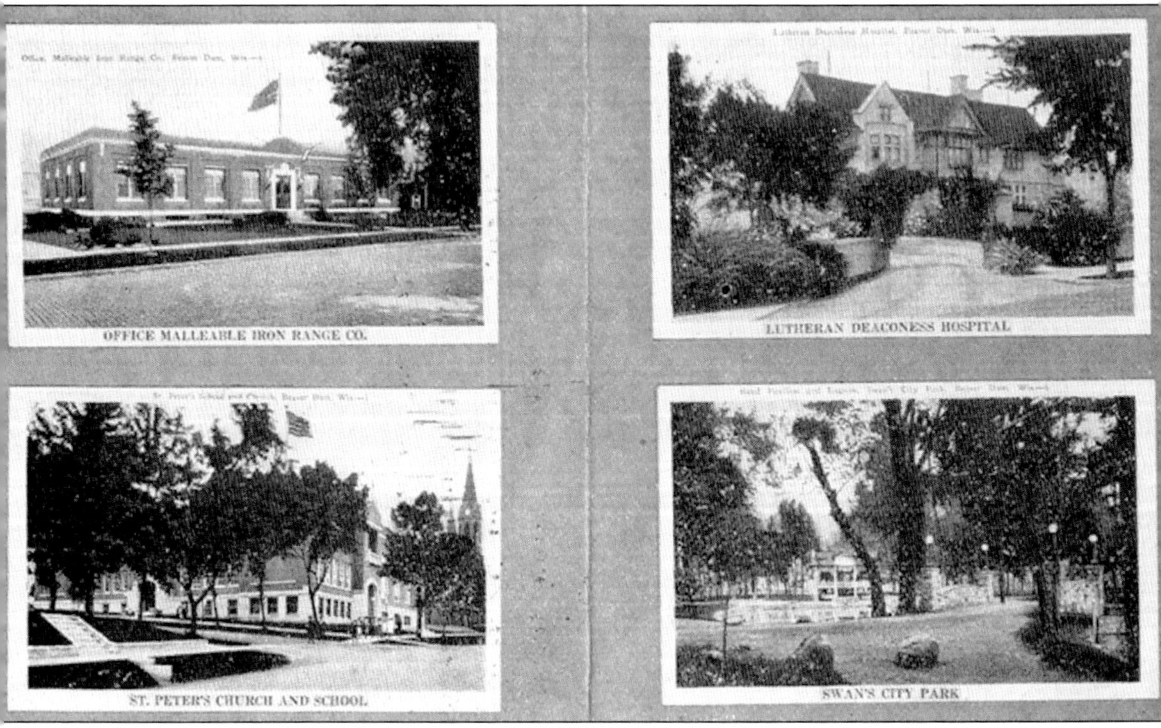

The Double-Sided Interior. The double-sided interior page of the souvenir postcard contained scenes of the Malleable Iron Range Company's office, the Lutheran Deaconess

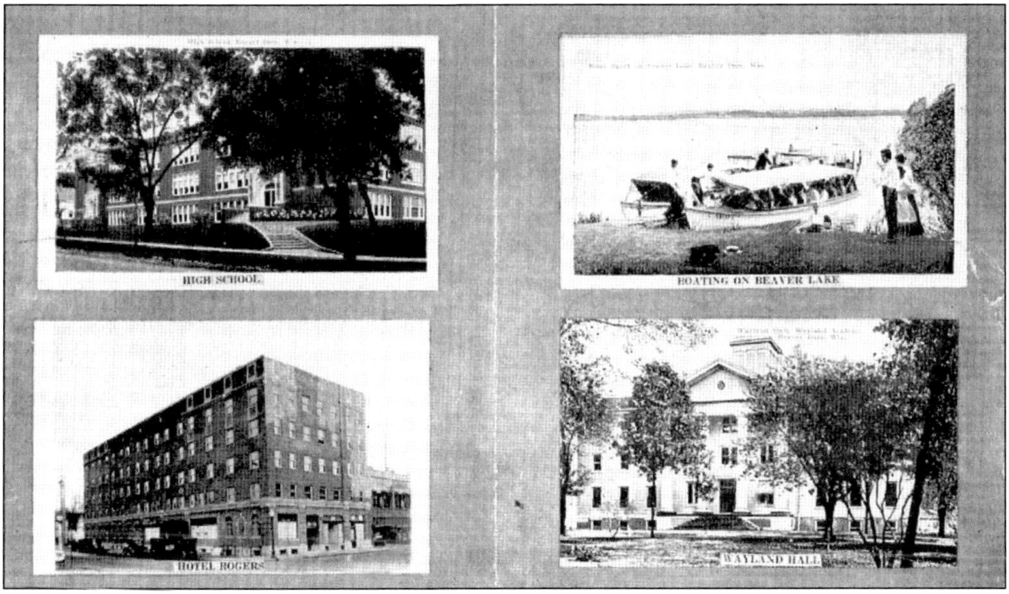

THE SINGLE-SIDED INTERIOR. The single-sided interior page of the souvenir postcard contained scenes of the high school, the lake boats, the Hotel Rogers, and Wayland

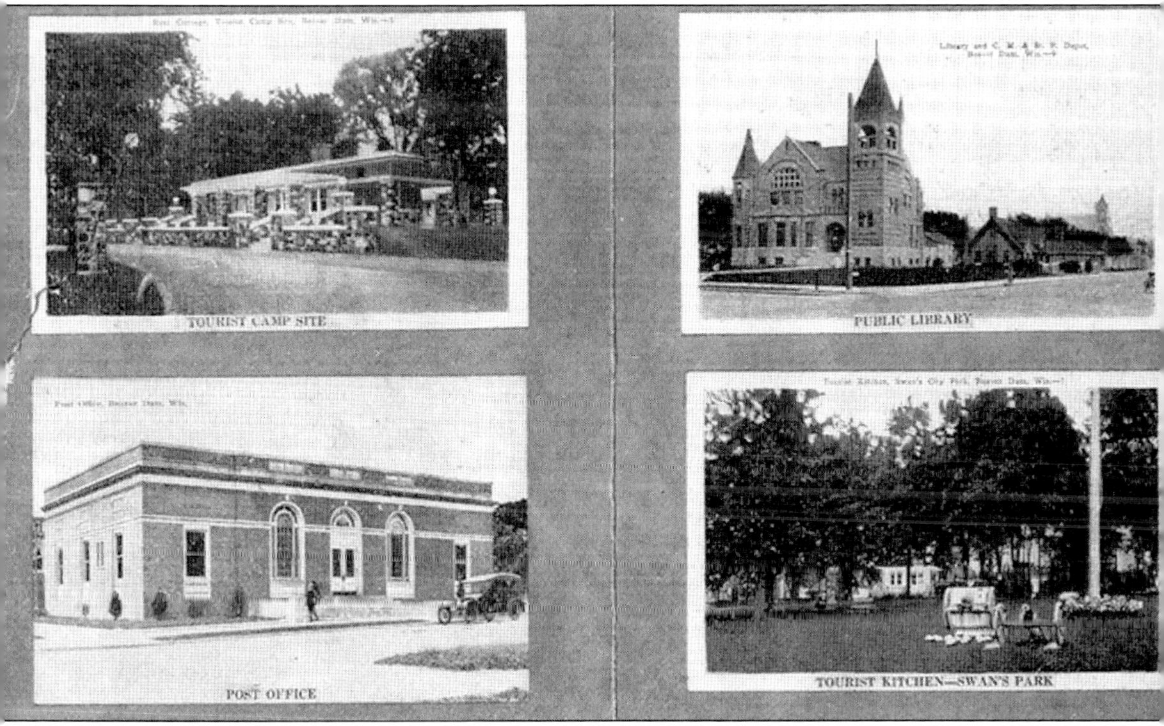

Hospital, St. Peter's School, Swan City Park, the Swan City Park tourist campsite, the Williams Free Library, the post office, and Swan City Park's tourist kitchen.

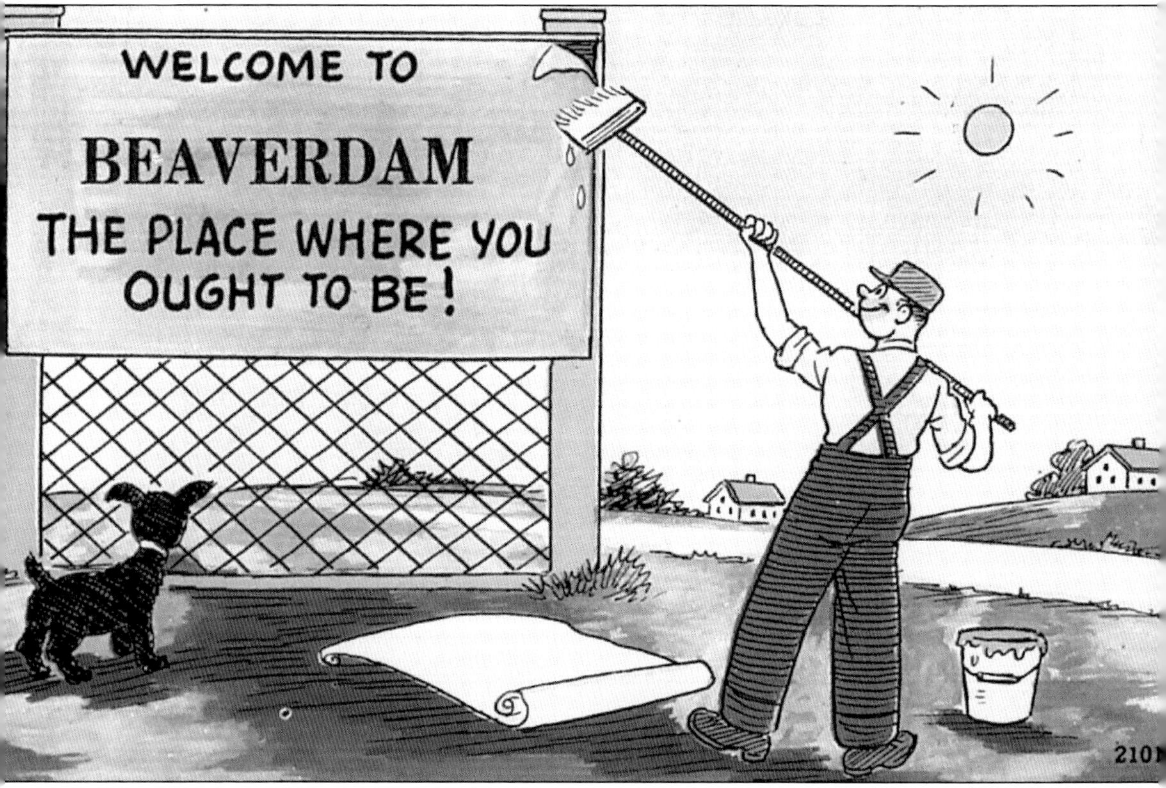

THE CARTOON POSTCARD. By the 1940s, the cartoon postcard had abandoned the "kewpie doll" illustration in favor of a drawing style more familiar to the readers of comic books or the Sunday funny pages.

Two

Downtown Where All the Lights Are Bright

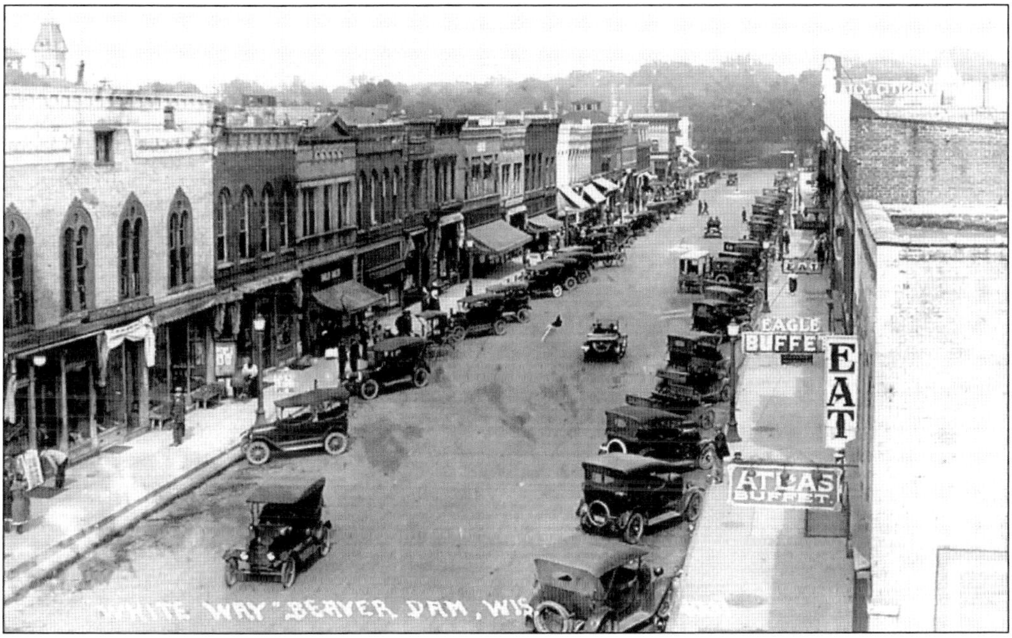

The Great White Way. In the second decade of the 20th century, every small city's downtown area boasted the title "Great White Way." The reference is to the first city street to be lit by electricity. The placement of the downtown lighting in Beaver Dam was funded by local merchants in 1911. The streetlights were then donated to the city.

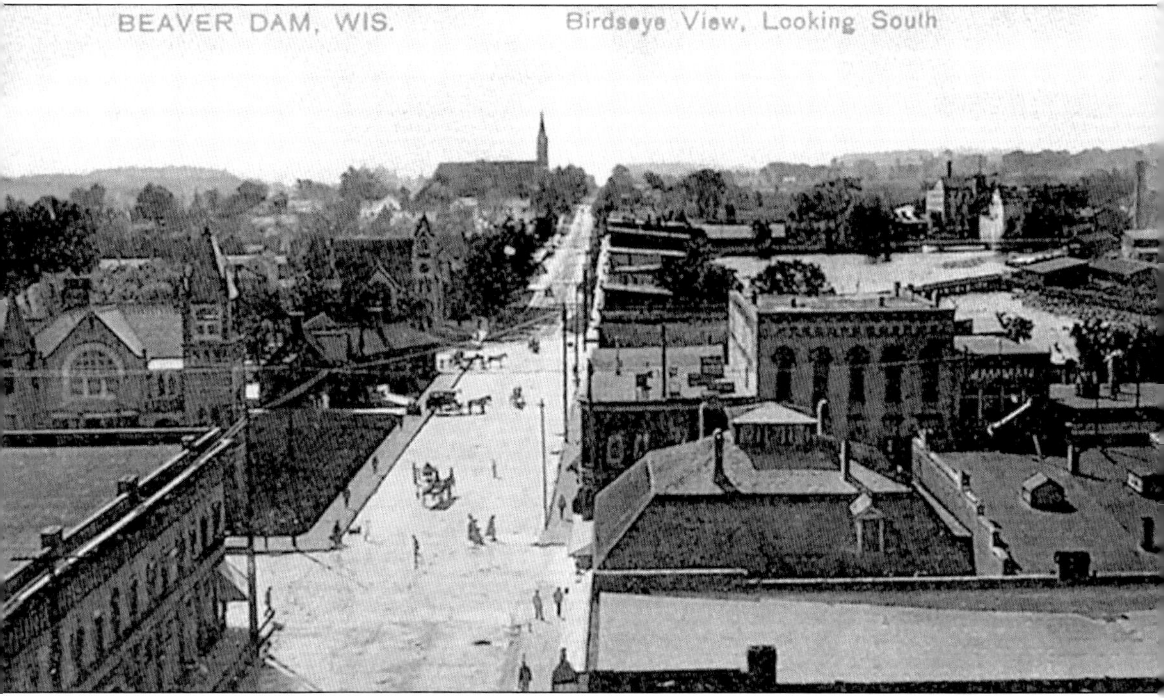

DOWNTOWN AT THE TURN OF THE CENTURY. This double-sided postcard presents an excellent bird's-eye view of the downtown and beyond around 1910. The photograph was taken from the bell tower of city hall. To the far left is the Williams Free Library and the Knights of

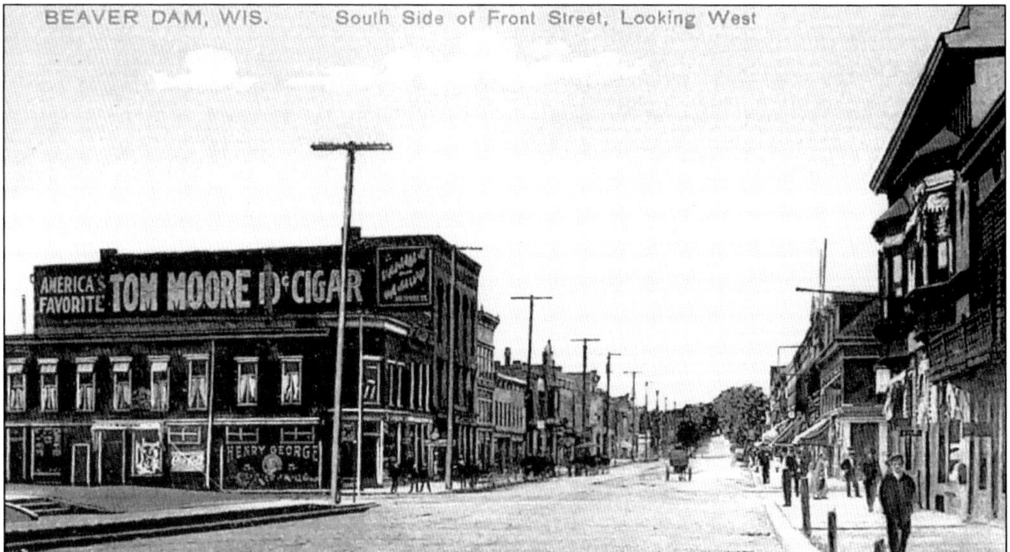

FRONT STREET, LOOKING WEST. At the turn of the century, the center of life in Beaver Dam was Front Street, which was comprised of dirt roads and board sidewalks. The telephone poles on the south side of the street had recently been installed. The C. W. Helbing drugstore and attorney George B. Swan occupied the southwest corner. On the northwest corner, in the Empire Block, Shepard and McFayden offered the finest in dry goods. At the northeast corner, on Park Avenue, O. Stein's newsstand offered the latest in books, magazines, and newspapers.

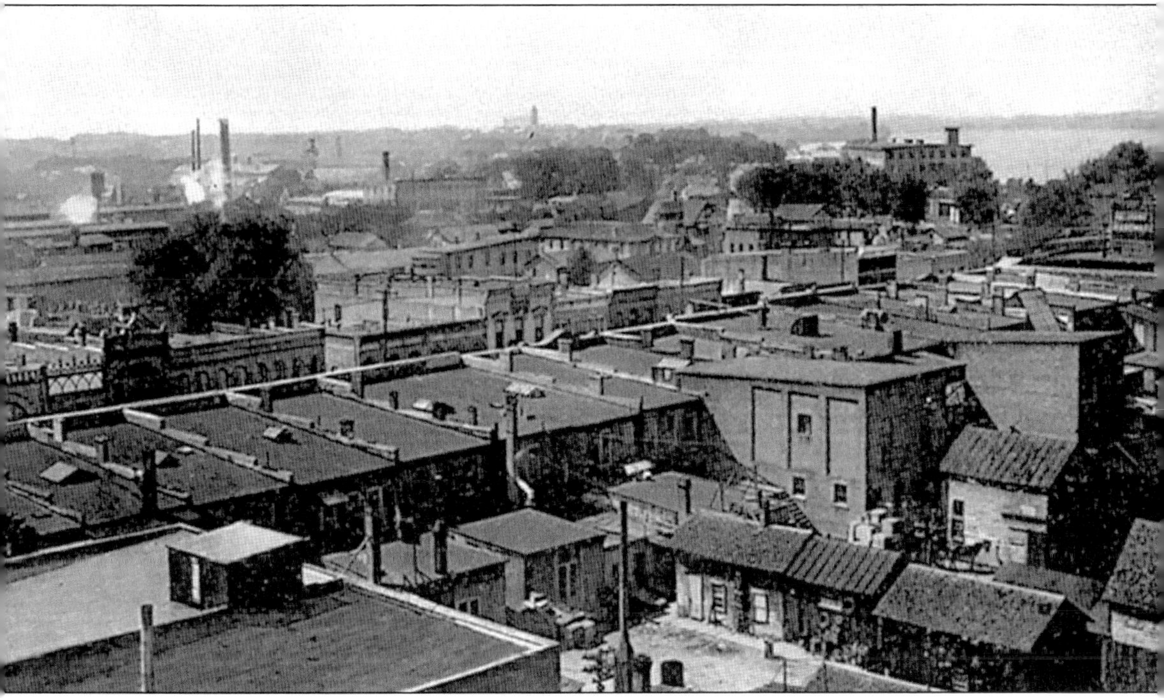

Columbus building (now Water Technologies). Farther south, St. Peter's Church can be seen in the background. Just south of Front Street is the millpond. To the far right is the cotton mill (now Weyco Shoe Factory).

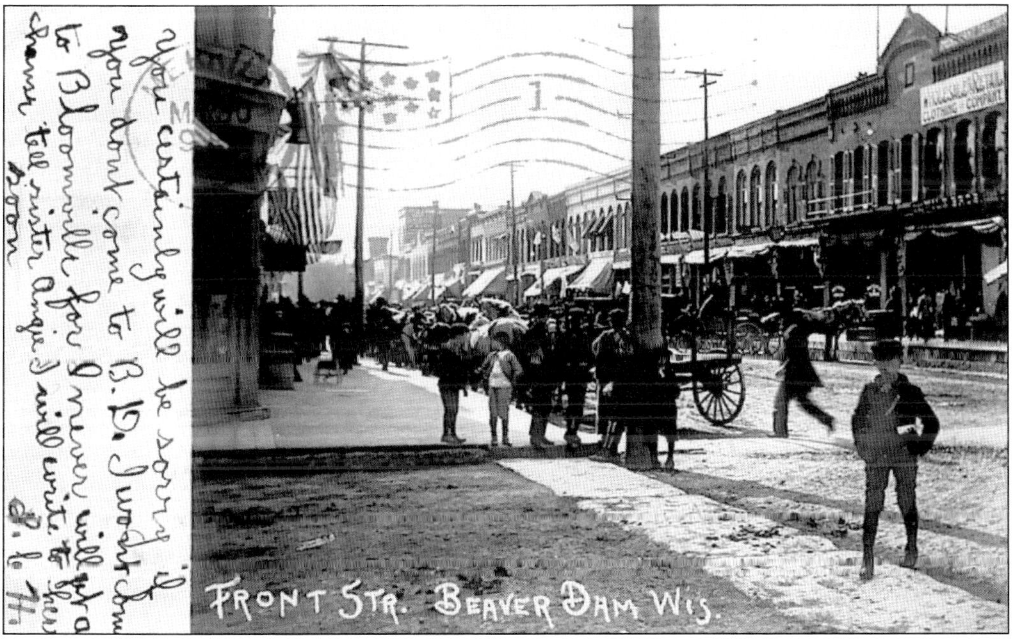

THE SOUTHWEST CORNER OF FRONT STREET. This view looks north from the southwest corner of Front and Spring Streets. The shops across the street include Wilk Brothers Clothiers, Thorp Jewelers, and the Hughes Drugstore.

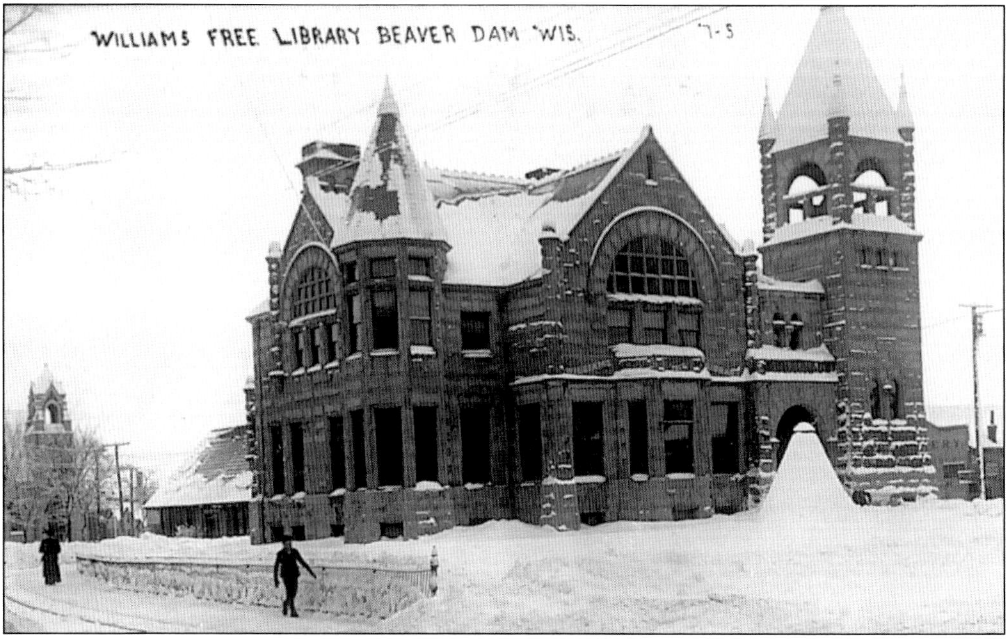

THE SOUTHEAST CORNER OF FRONT STREET. One of the great landmarks of Beaver Dam is the Williams Free Library. The building, completed in 1891, was donated to the community by J. J. Williams. In the background of this photograph can be seen a portion of Wade's Livery Stable, which was located on Spring Street. The library was placed on the National Register of Historic Places in 1974. It is the current home of the Dodge County Historical Society.

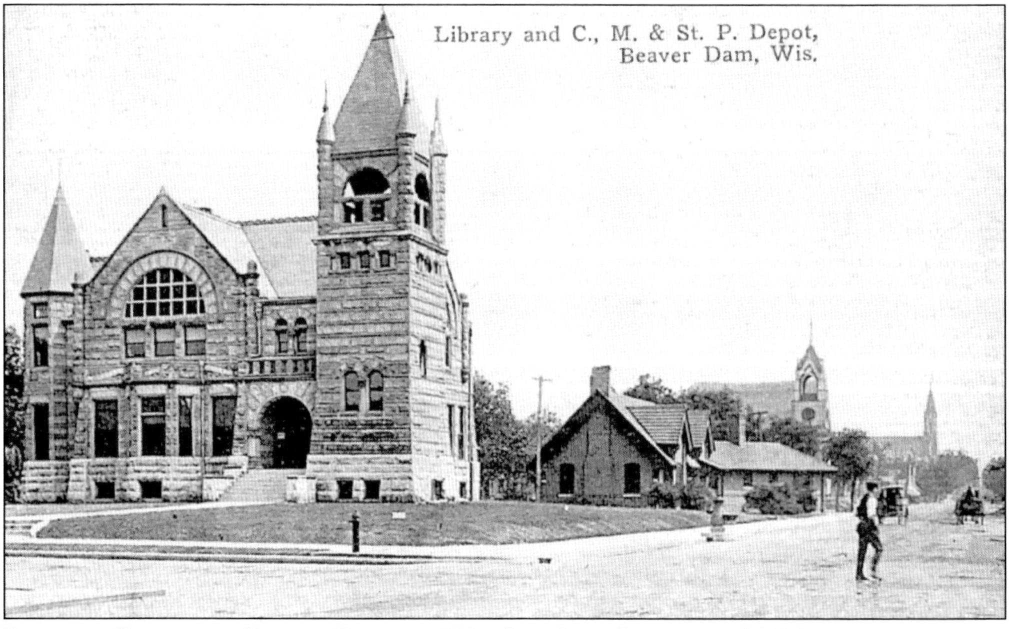

LOOKING SOUTH ON SPRING STREET. This view dates from around 1900.

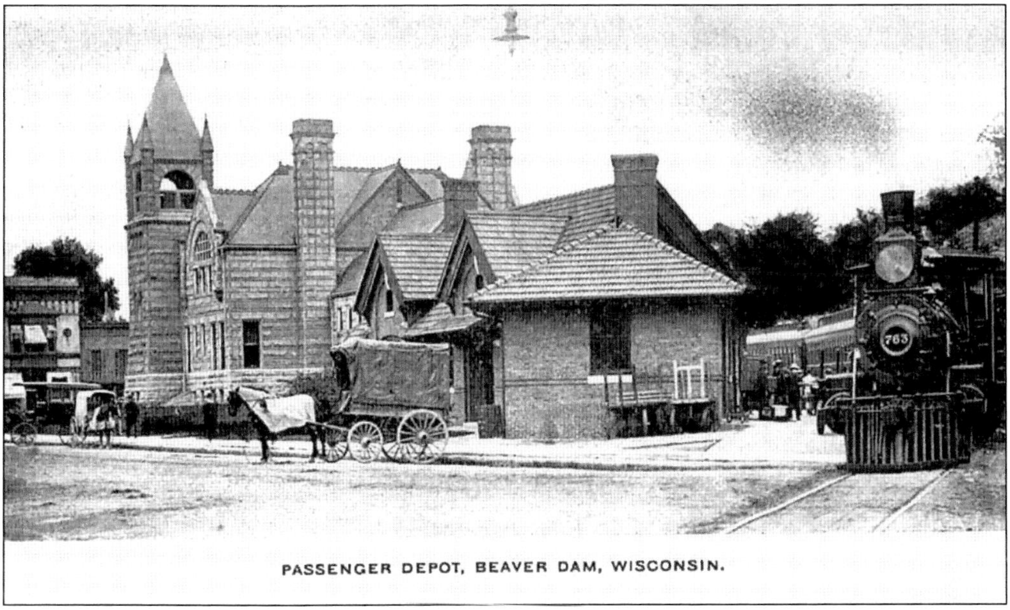

THE PASSENGER DEPOT. In 1872, the LaCrosse & Milwaukee Railroad laid the "fryin' pan" loop into downtown Beaver Dam. The first downtown depot, made of white brick, was replaced in 1900 by the above depot.

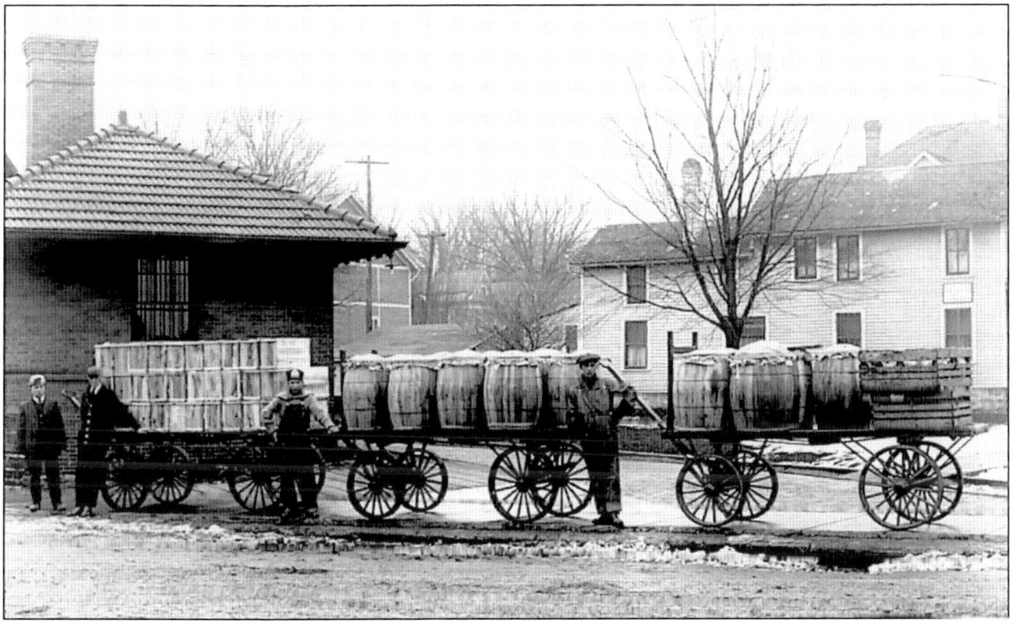

UNLOADING FOR DOWNTOWN STORES. From the turn of the century through the 1920s, nearly all of the merchandise shipped to Beaver Dam businesses was delivered by rail. As motor vehicles became more reliable, passenger and freight service declined. The depot closed in the early 1950s. The city took the title to the property, leasing it first to the Dodge County Historical Society in 1953 and then to the chamber of commerce in 1983.

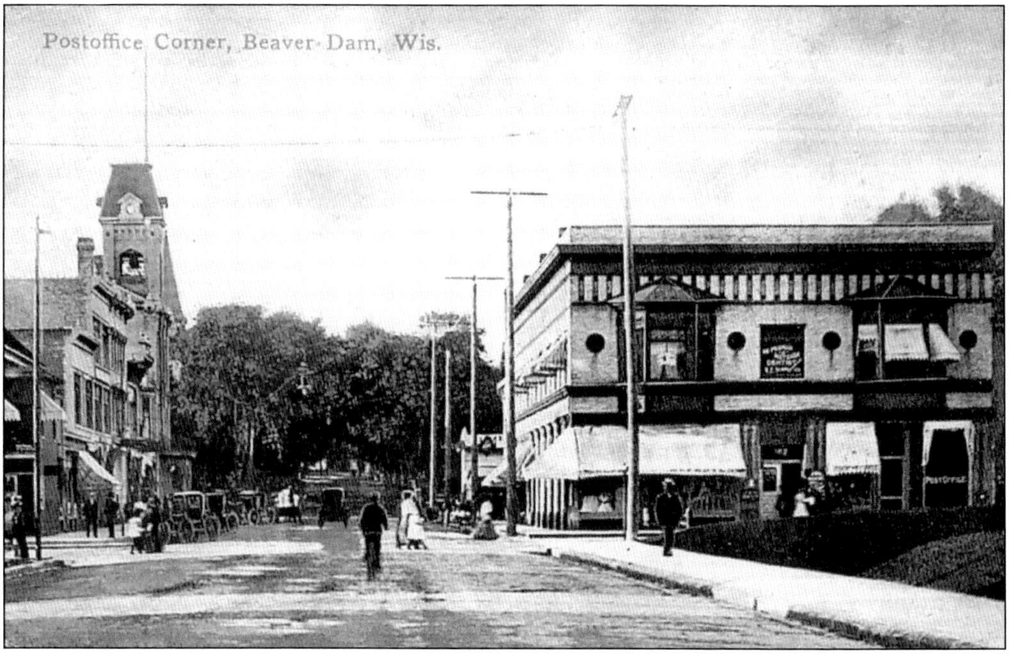

The Post Office Corner, c. 1910. In the early days of Beaver Dam, the post office was located wherever the current postmaster did business. This was the city's eighth post office.

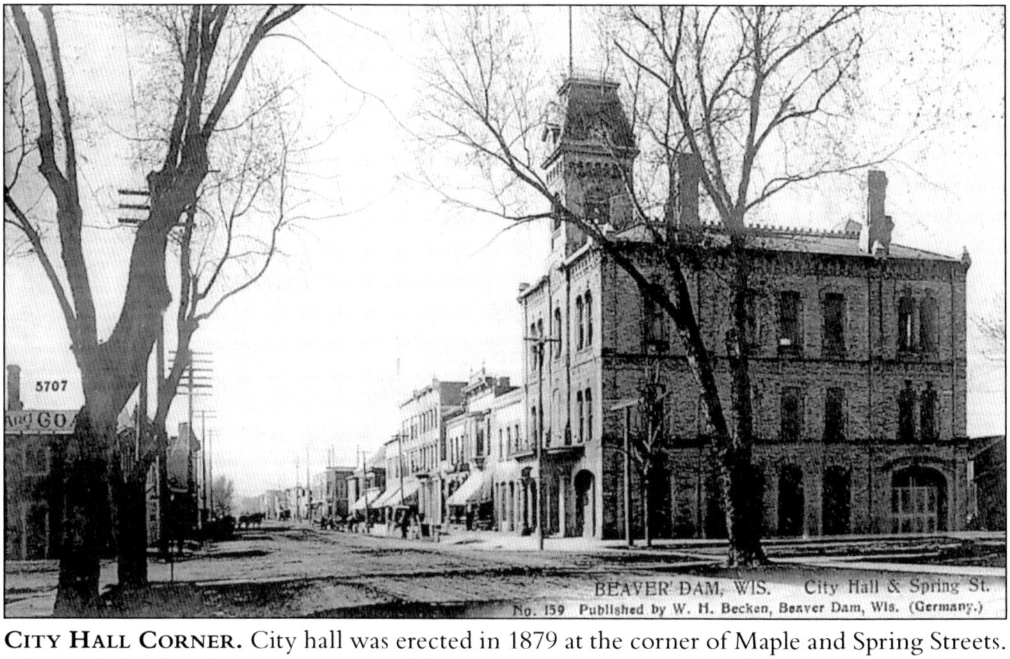

City Hall Corner. City hall was erected in 1879 at the corner of Maple and Spring Streets. It housed the fire department, city offices, and a gymnasium. In 1957, a new city hall was constructed at the site of the old Washington Grade School on South Lincoln Avenue. The old city hall was sold to Farmer's State Bank.

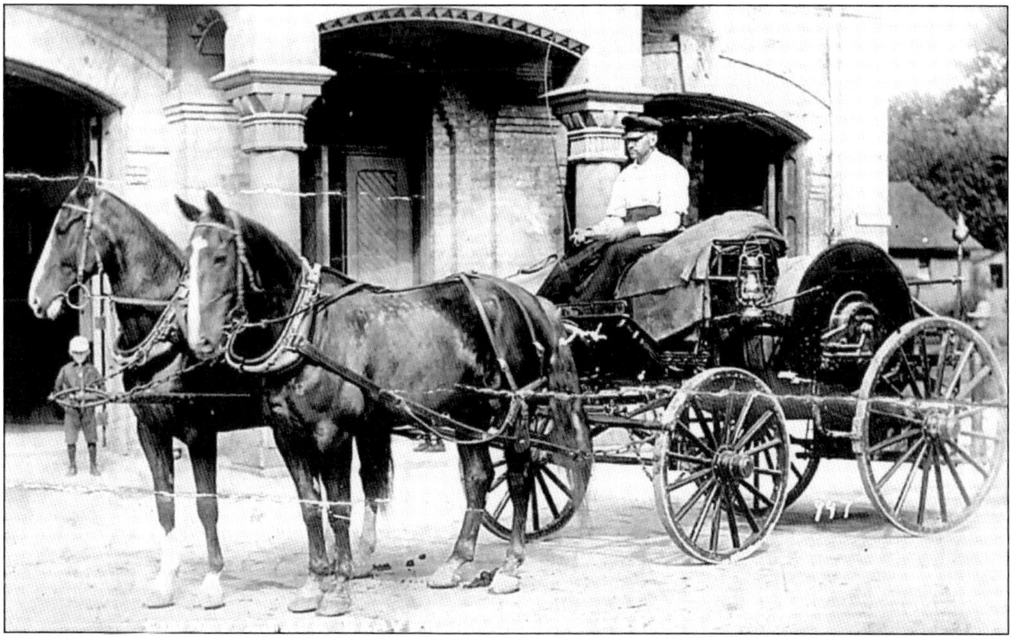

The Fire Department. The fire department was organized in 1866, and the first firehouse was located on Madison Street. The fire department relocated to the first floor of the new city hall when it was completed in 1879. Pictured above is driver George Sage with horses Sidney and Jack in harness. Sage's son Merlin stands at the station door.

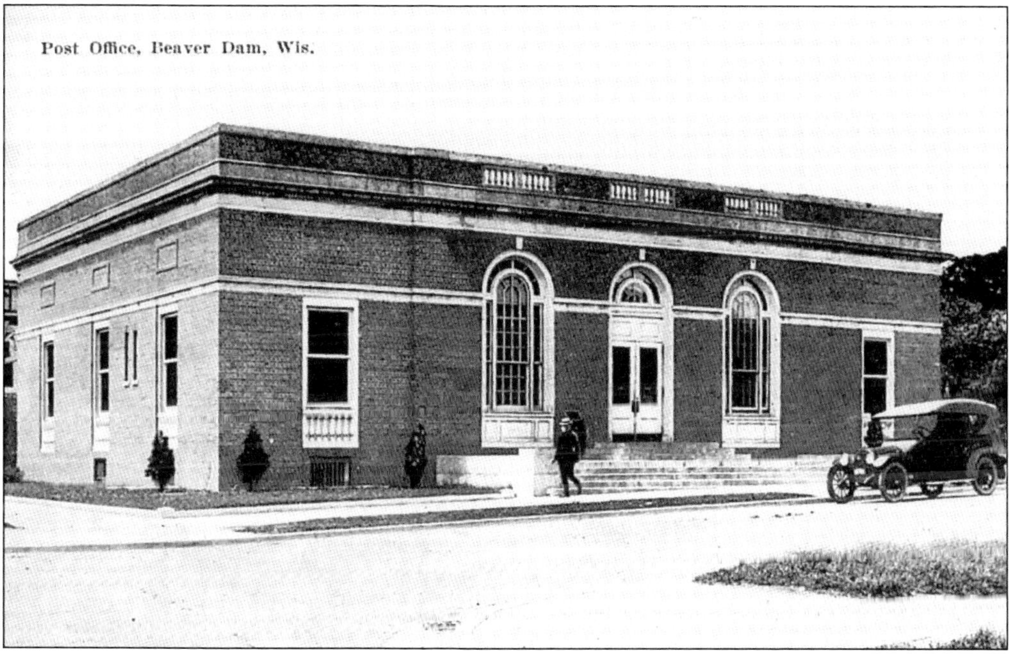

The New Post Office. Built in 1918, this government-owned building became the first permanent home of the post office in Beaver Dam. Due to parking limitations and size constraints, this building was sold to Farmer's State Bank in 1964. In that same year, the government leased the current post office site at 411 North Spring Street.

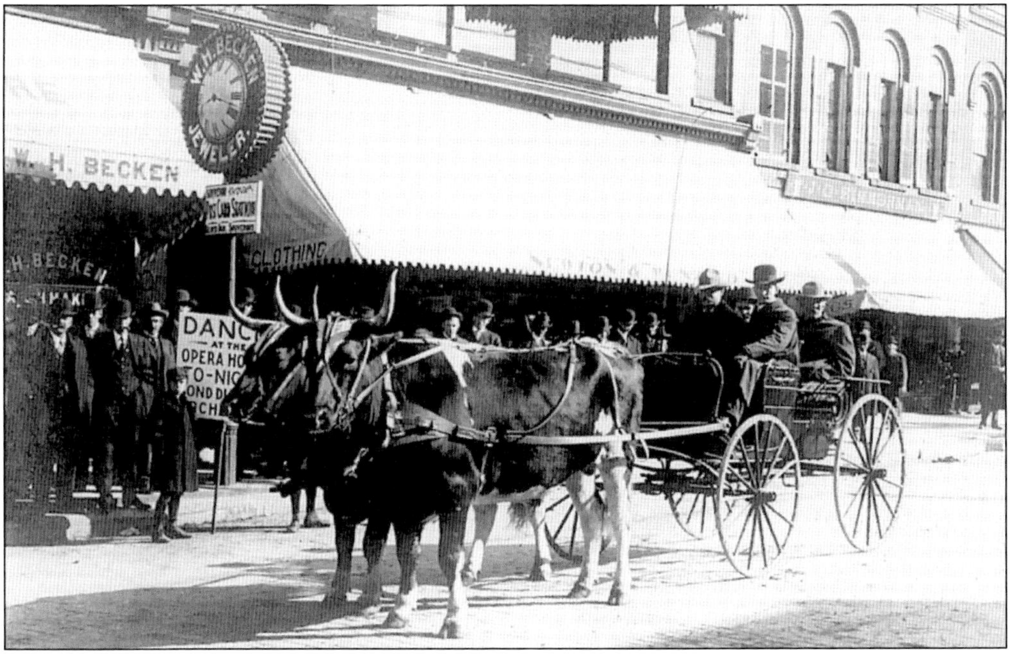

The W. H. Becken Store. W. H. Becken was born in 1869. After holding numerous occupations throughout Wisconsin, Becken settled in Beaver Dam, opening a jewelry store here in 1903. In 1909, he sold his jewelry business to enter into the sale of general merchandise, and he continued in that endeavor into the late 1920s. The caption for this photograph reads, "The Last Oxen Team in the United States."

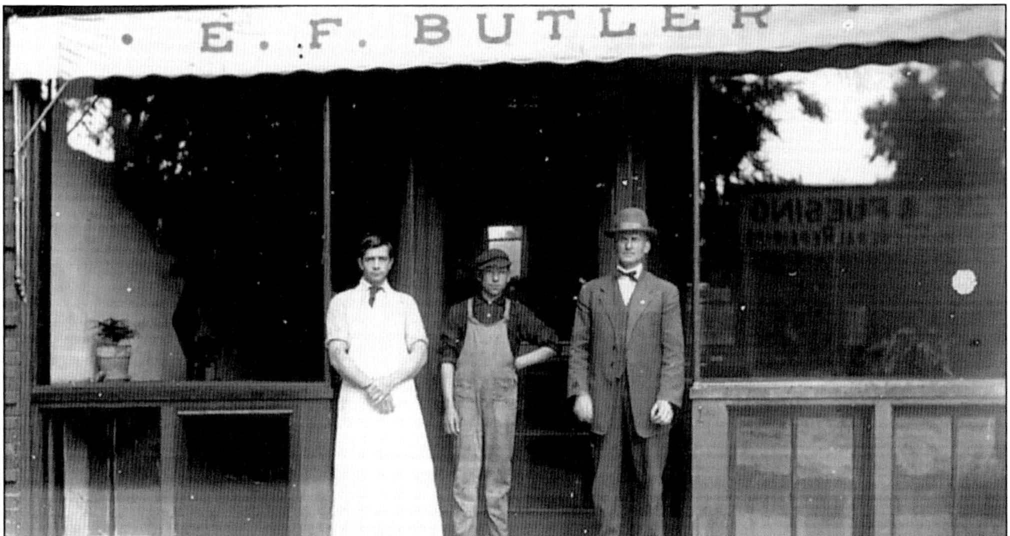

Eugene F. Butler. Eugene Butler ran a meat shop and delivery service out of a storefront located at 207 North Spring Street in the second decade of the 20th century. In the 1930s, Butler teamed with C. F. Daniel to run a coal and wood delivery service out of a Rowell Street building. Pictured are, from left to right, Mike Roedl, Alois Ledworowski, and Eugene F. Butler.

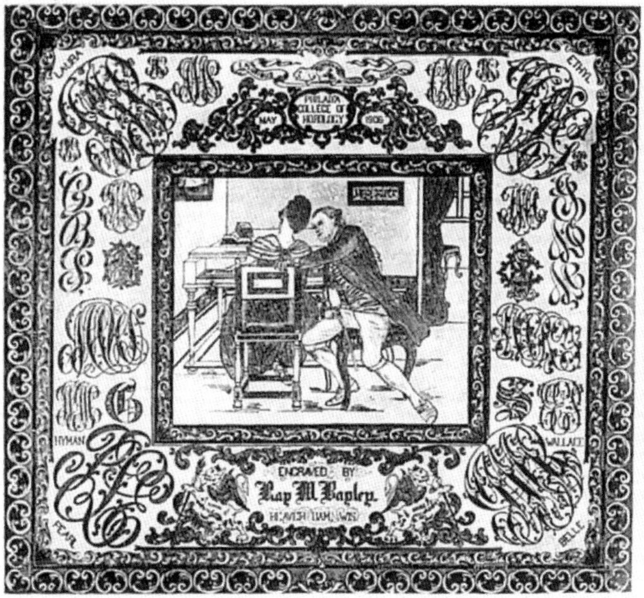

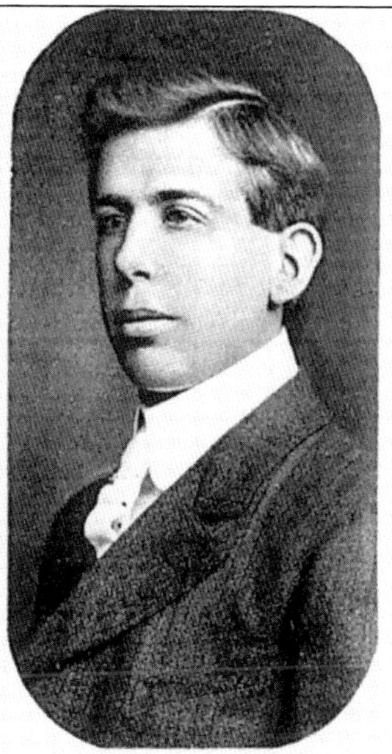

R. W. Bayley and Company. Ray W Bayley was born and raised in the Beaver Dam area. In 1905, he traveled to Philadelphia to apprentice as a jeweler. It was there that he became proficient in watchmaking. In 1906, he returned to Beaver Dam and worked under W. H. Becken. In 1911, he took over the business, expanding in the 1920s into optometry. He switched solely to the practice of optometry in the 1930s, and he continued in that profession into the 1960s.

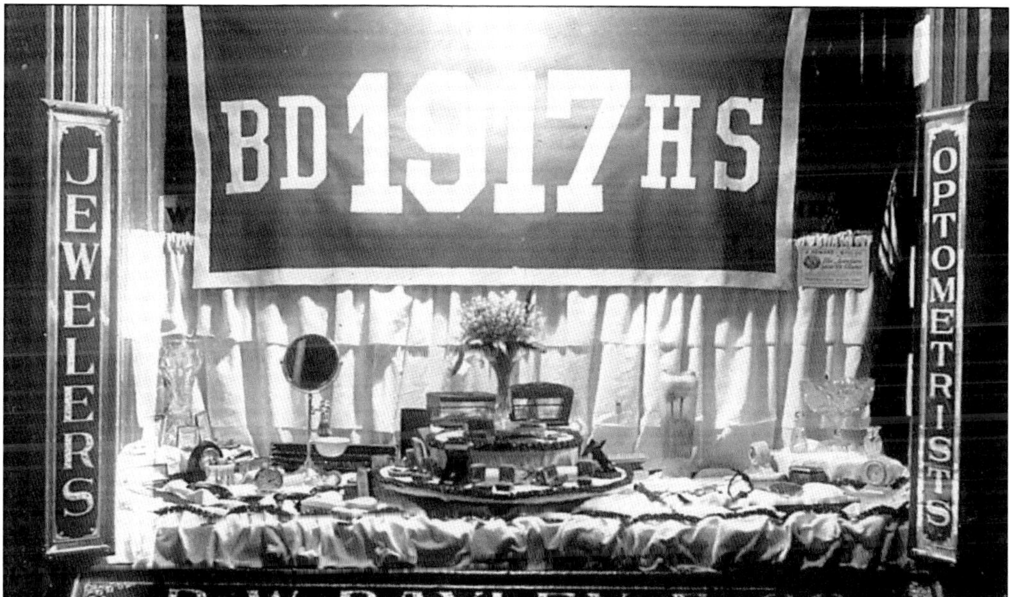

The Bayley Storefront. The Bayley store was located at 134 Front Street. This window display congratulates the Beaver Dam High School graduates of 1917 and offers suggestions for graduation presents, including watches, razors, and other toiletries.

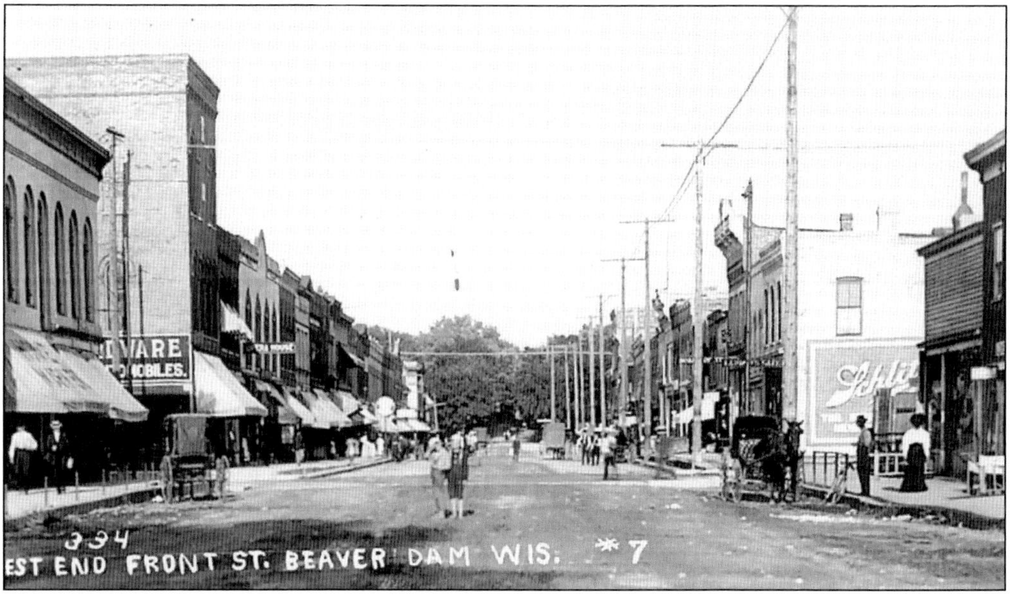

FRONT STREET, LOOKING EAST, C. 1900. Looking east from the corner of Front and Center Streets, this turn-of-the-century photograph reminds us of life prior to the horseless carriage. Located on the northeast corner is Rassman Hardware, and one building down is the Opera House. On the northwest corner is McLaughlin Groceries, and one door west is Gruenberg and Briese Furniture. At the southeast corner is the J. Good Saloon and Café.

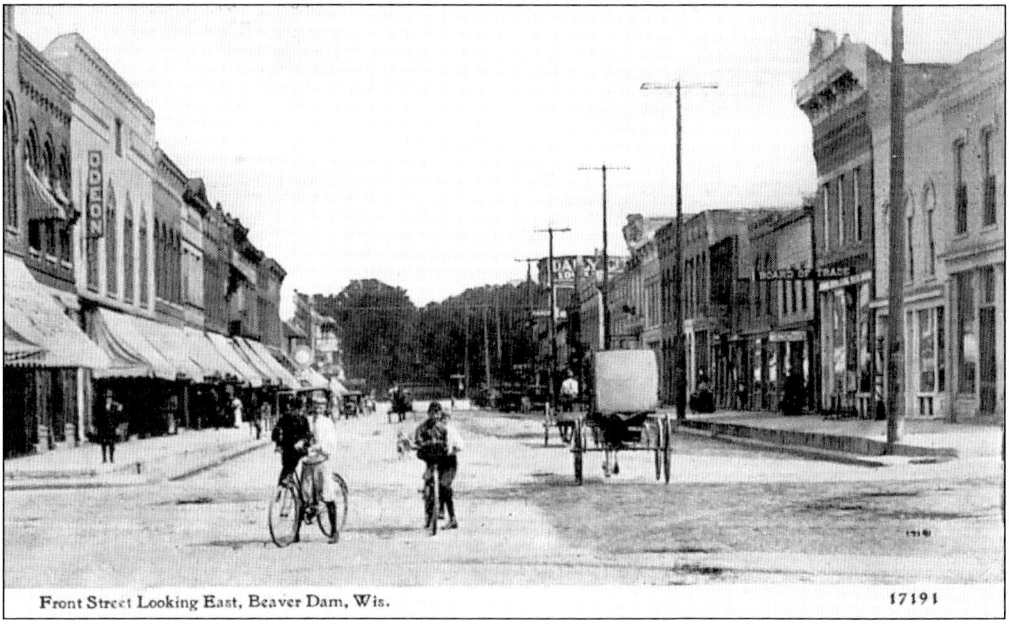

FRONT STREET, LOOKING EAST, C. 1910. A decade later, the opera house has been replaced with the Odeon Nickelodeon. The new popularity of the bicycle is in great evidence, and one automobile can be glimpsed in the distance. Halfway down the south side of the road, at 135 Front Street, stands McKinstry's Furniture. This landmark has stood the test of time, successfully transacting business for 147 years and counting, making it the oldest family-run furniture store in Wisconsin.

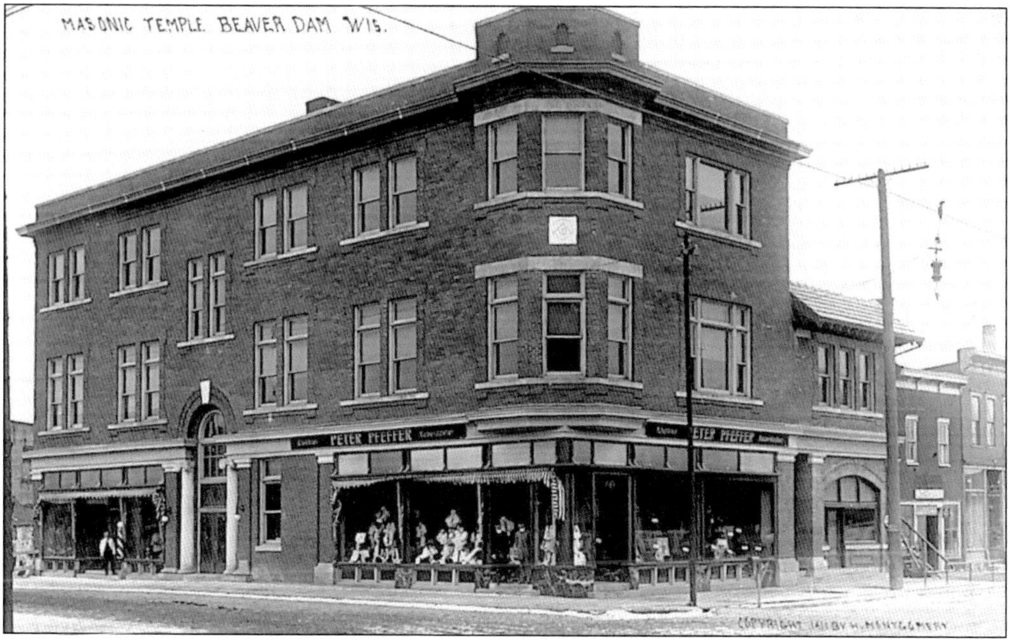

THE MASONIC TEMPLE. The Masonic Order of Beaver Dam, Dodge County Lodge No. 72, was chartered in 1856 with a membership of 20. It was housed in the Voreck Building prior to the 1863 fire and later occupied portions of the Van Kanel and Schultz Brothers Buildings. On March 25, 1911, the order relocated to its own building, pictured here.

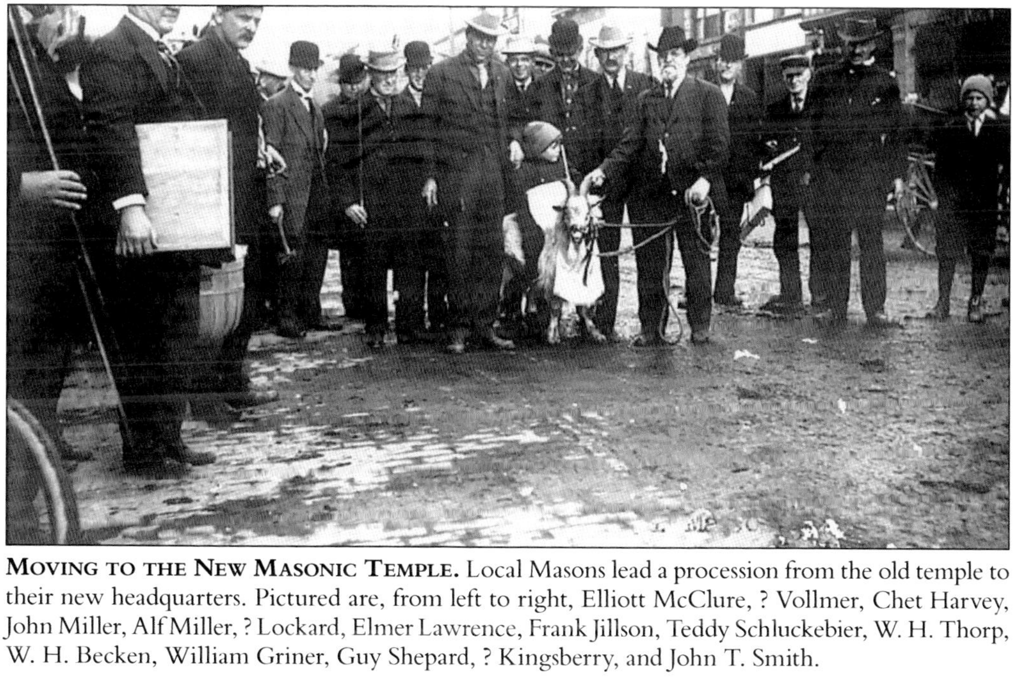

MOVING TO THE NEW MASONIC TEMPLE. Local Masons lead a procession from the old temple to their new headquarters. Pictured are, from left to right, Elliott McClure, ? Vollmer, Chet Harvey, John Miller, Alf Miller, ? Lockard, Elmer Lawrence, Frank Jillson, Teddy Schluckebier, W. H. Thorp, W. H. Becken, William Griner, Guy Shepard, ? Kingsberry, and John T. Smith.

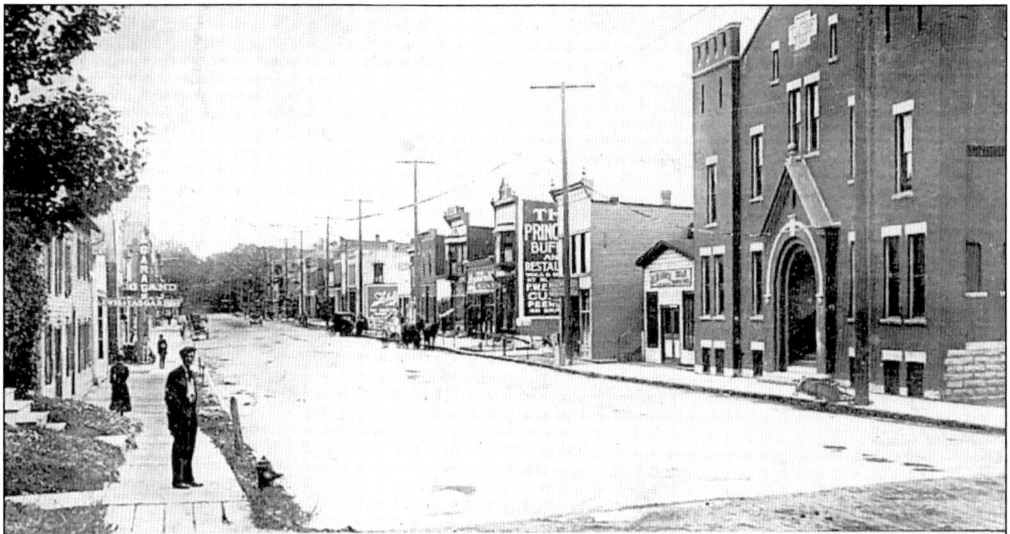

Front Street, Looking East, Beaver Dam, Wis.

FRONT STREET, LOOKING EAST FROM BEAVER STREET. Taken between 1908 and 1911, prior to the construction of the Davison Theatre, this photograph shows a city's infrastructure in transition. The bricking of Front Street, begun in 1902, did not yet extend past Beaver Street. While the newly built armory is fronted by cement sidewalks, the sidewalks on the north side of the street are still composed of wood planks. Businesses on the south side of the street include Harder Cigar (next to the armory) and the Princess Buffet (down the block). On the north side of the street, one could shop for home furnishings at Lewis and Yasger Furniture or have that new motorcar serviced at the Silo Garage.

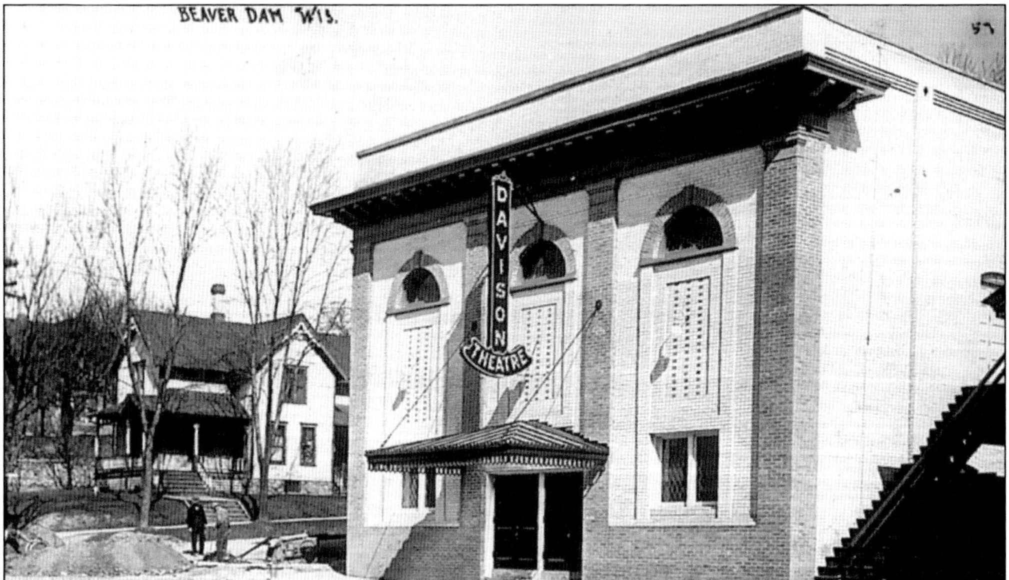

THE DAVISON THEATRE. The Davison Theatre, built in 1911 at the corner of Front and Beaver Streets, offered the latest in live professional and community theater. A competitor of the Concert Hall and the Music Hall, the Davison also offered silent motion pictures. In 1948, the Davison Theatre changed hands and was renamed the Wisconsin Theatre.

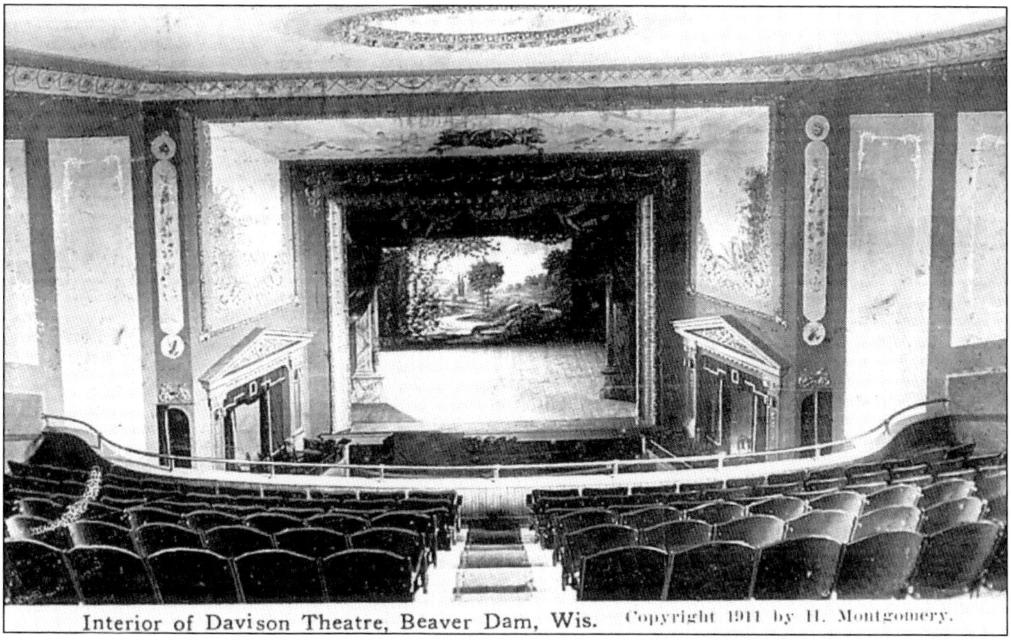

THE DAVISON STAGE AND AUDITORIUM. These views offer a glimpse of the Davison interior.

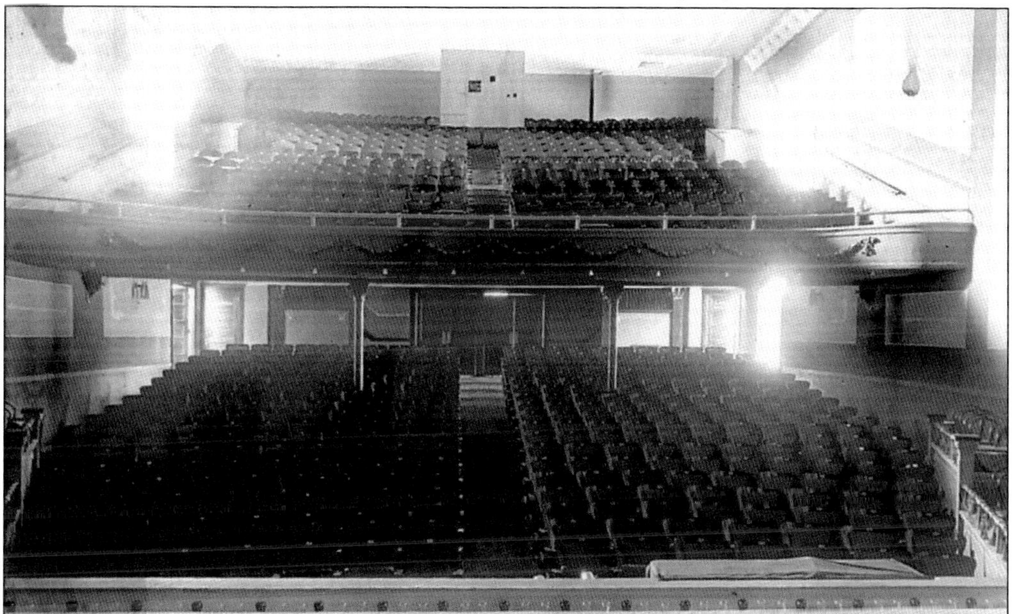

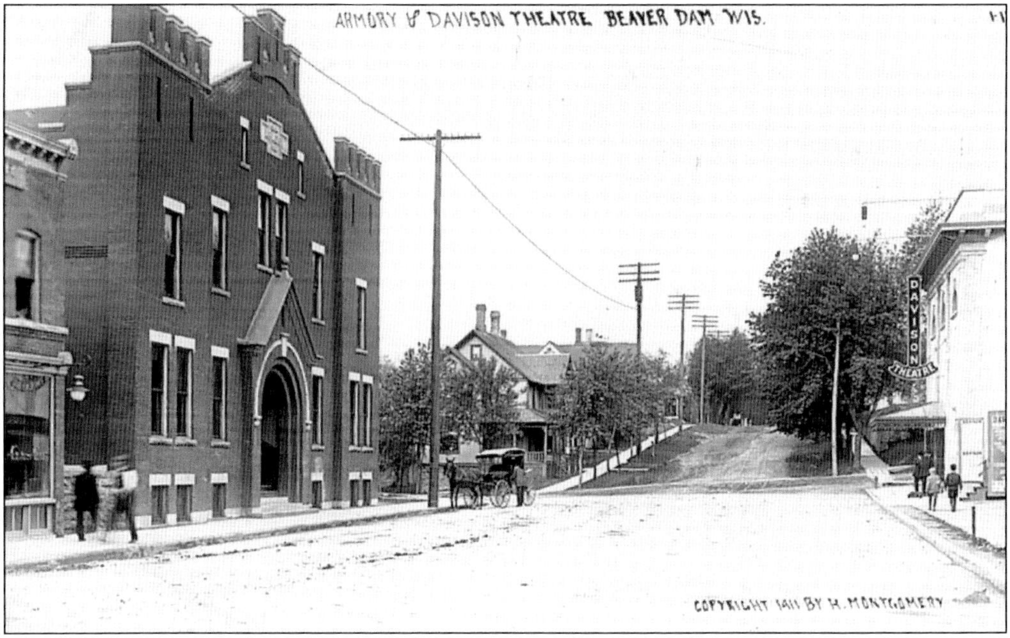

THE ARMORY. In 1908, the Beaver Dam Armory found a home on the southeast corner of Front and Beaver Streets. The home of Company K, the armory was a place for meetings, drilling, and dancing.

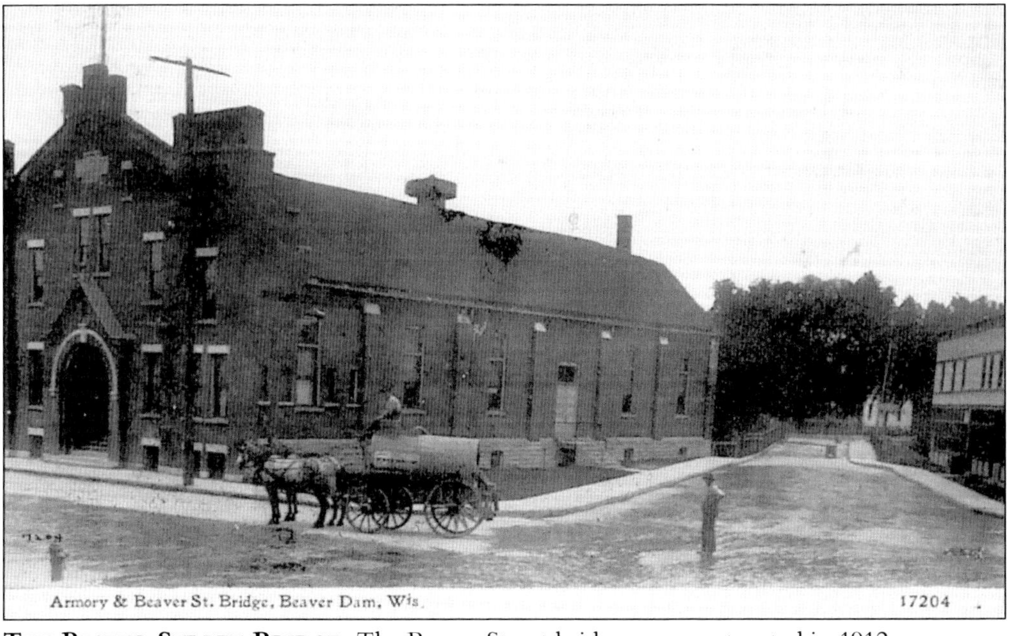

THE BEAVER STREET BRIDGE. The Beaver Street bridge was constructed in 1912.

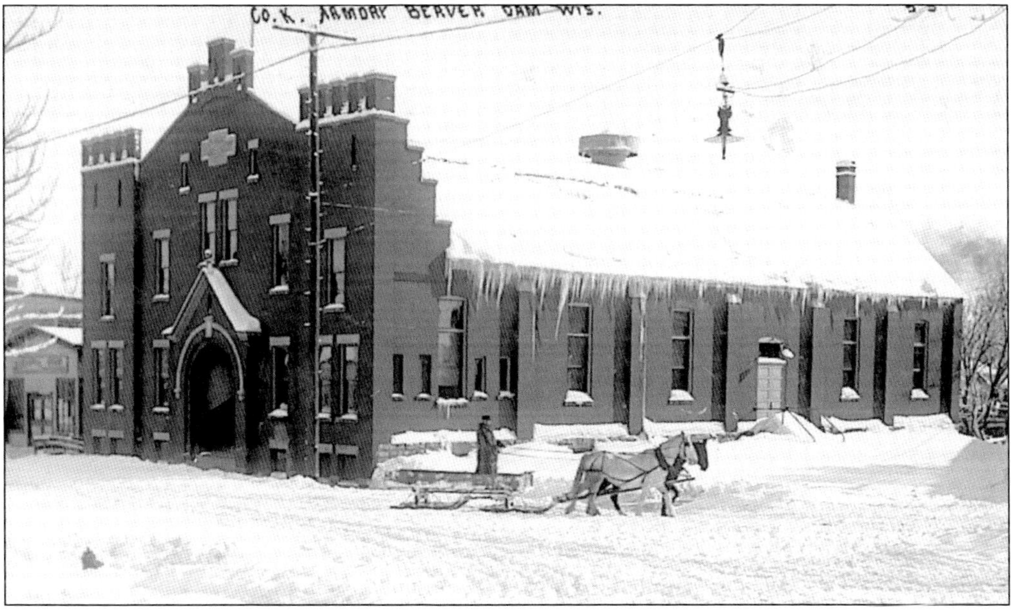

THE ARMORY IN WINTERTIME, *c.* 1910. The armory had served Beaver Dam's National Guard unit for 63 years when, in 1971, this structure was razed to be replaced by a Kentucky Fried Chicken restaurant.

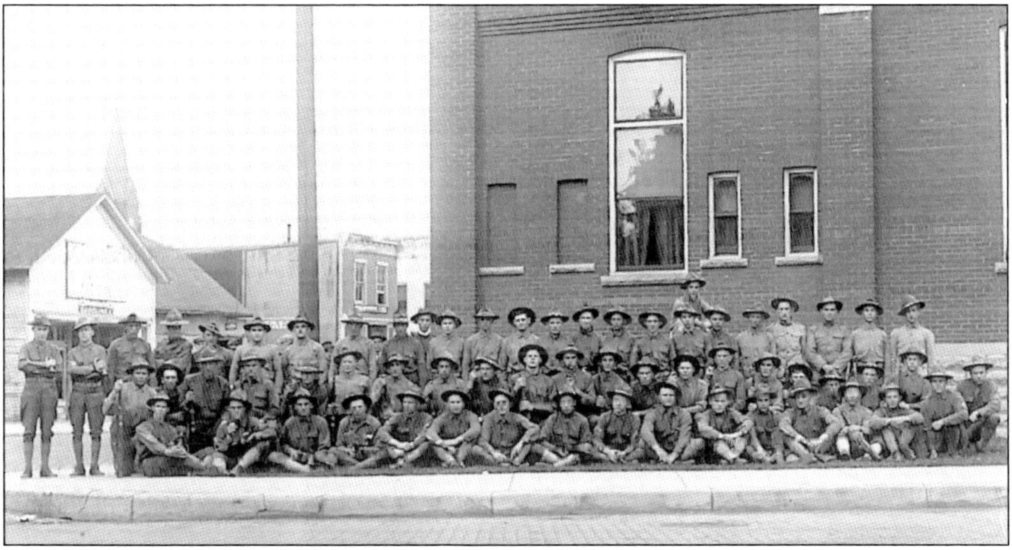

COMPANY K. This photograph was taken just prior to World War I. Of the men pictured, 11 made the ultimate sacrifice, and two-thirds were wounded in action.

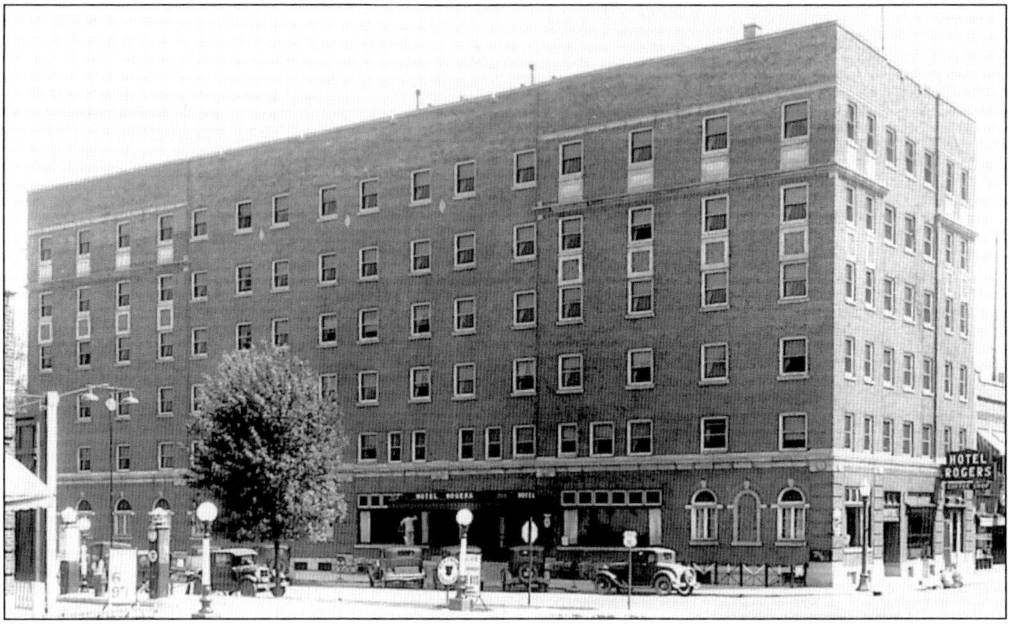

THE HOTEL ROGERS. In 1927, F. W. Rogers, along with other prominent Beaver Dam businessmen, sought to create a state-of-the-art hotel that would serve as a destination for traveling salesmen and tourists as well as a site for conventions. The result was the Hotel Rogers. Initially five floors tall, the hotel added a sixth floor in 1931, expanding to 140 rooms. The Hotel Rogers was placed on the National Register of Historic Places in 1989.

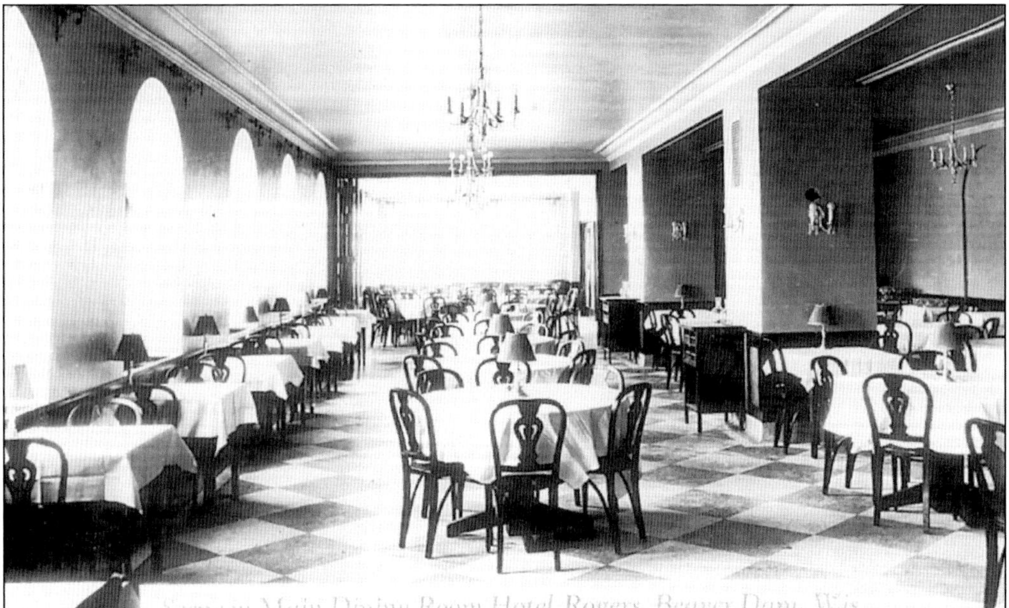

THE HOTEL ROGERS' DINING ROOM. One of the main attractions of the hotel was its elegant restaurant. The gourmet food attracted regular patrons from as far away as Milwaukee, Chicago, and Minneapolis.

THE HOTEL MANLY, 1920S. Beaver Dam's premier hotel prior to the construction of the Hotel Rogers was the Hotel Manly, located at 110–112 North Spring Street. The hotel was originally built in 1866, and additions followed in 1904 and 1913. The Hotel Manly was known by many names, including the Schulze Hotel and the Beaver Hotel. The building was razed in 1966.

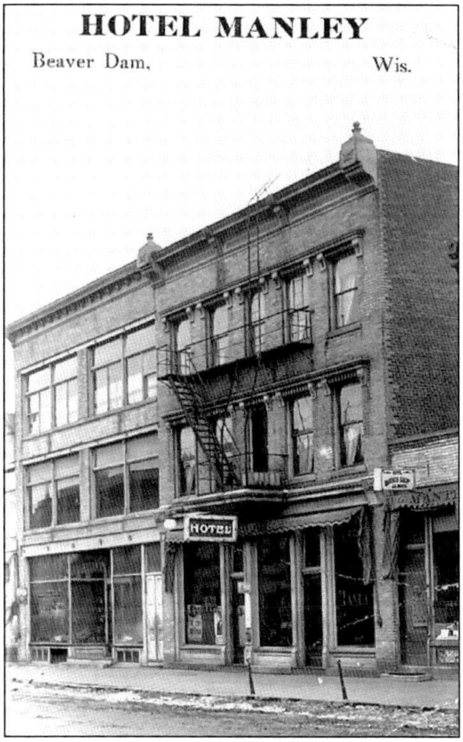

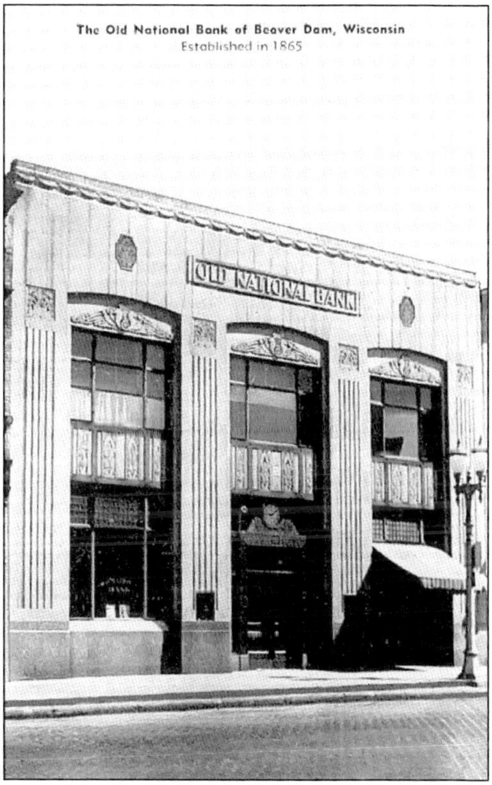

THE OLD NATIONAL BANK, 1930S. The only major private building erected in the 1930s was the Old National Bank, located at 138 Front Street (the former site of Jacob Brower's log cabin). The bank called this site home as early as 1865; this building was constructed in 1931. The bank relocated to Park Avenue in 1965.

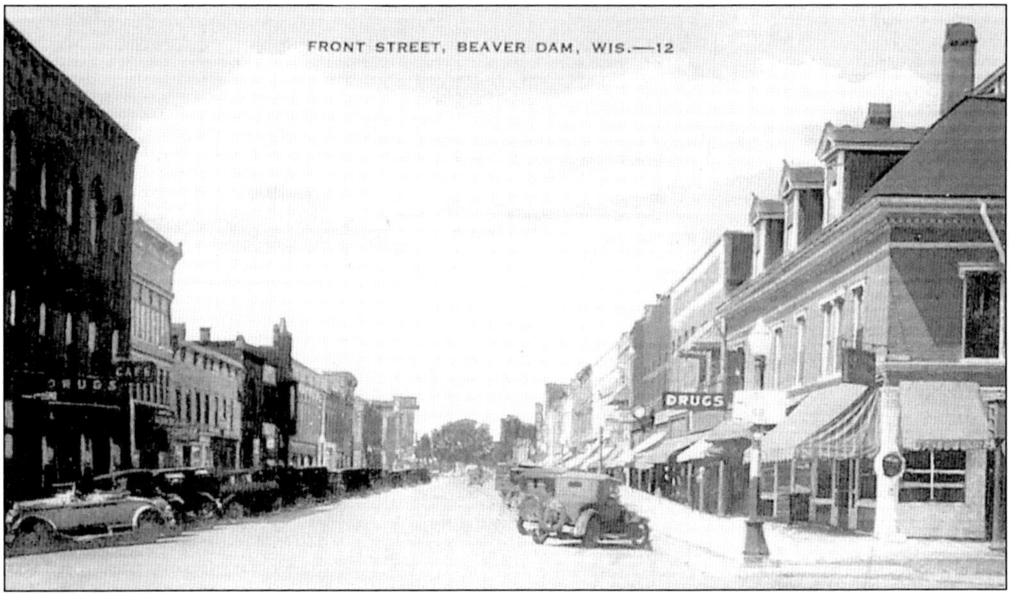

FRONT STREET, LOOKING WEST, 1930S. The downtown was dominated by the automobile in the 1930s. The streets were completely bricked, and all of the sidewalks were cemented. The stores on the north side of the street included Liebigs (on the corner), Langmack's Drugstore, and Maier and Zahn Clothiers (which replaced Wilks Brothers). Across the street, Roger's Pharmacy continues as a Front Street mainstay.

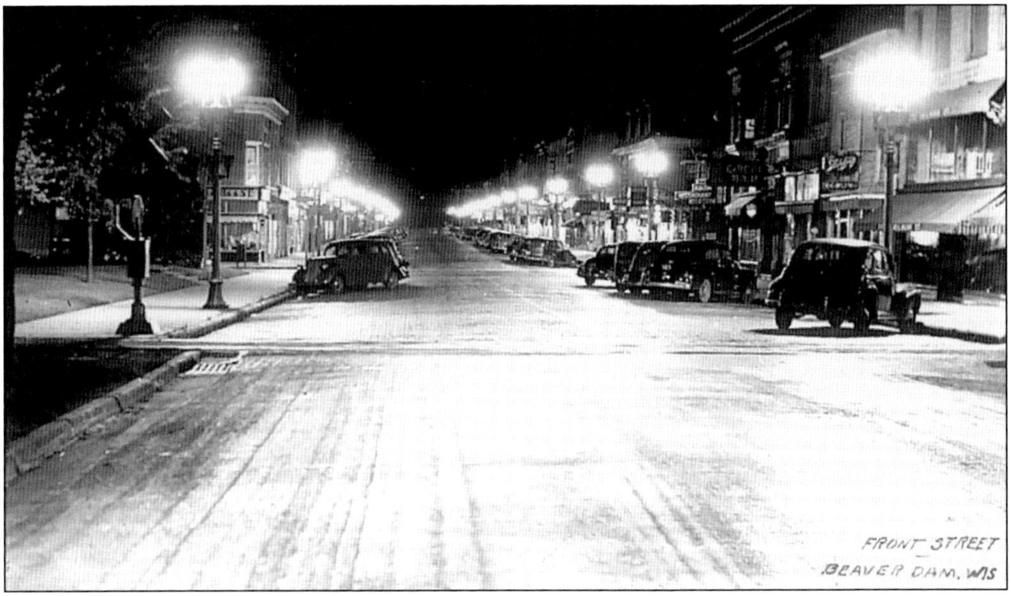

FRONT STREET, LOOKING WEST, 1940S. In the 1940s, Front Street underwent further improvement. The streets were paved with concrete, and traffic lights were put in place. The single-bulb streetlights had been replaced by double-bulb streetlights in 1939. The stores fronting the north side of Park Avenue, proceeding eastward from the corner, are Aldon Restaurant, the Circus Bar, Maier's Bowling Alley, and Briese Furniture. At the northwest corner of Front and Spring Streets are Beaver Floral and Martin's Red Goose Shoe Store. The southwest corner is occupied by the O.A. Haase Shoe Store.

THE MODEL. Located at 126 Front Street, Elliott's Model was advertised in 1941 as "one of the finest shops in Wisconsin catering to the Women and Miss who want to buy the newest in apparel."

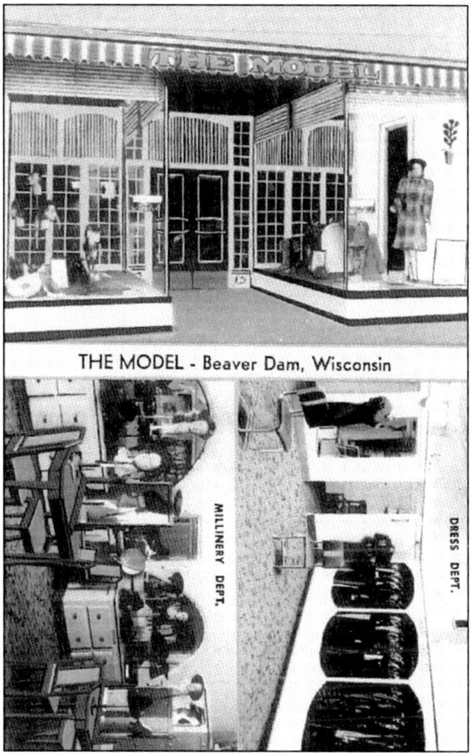

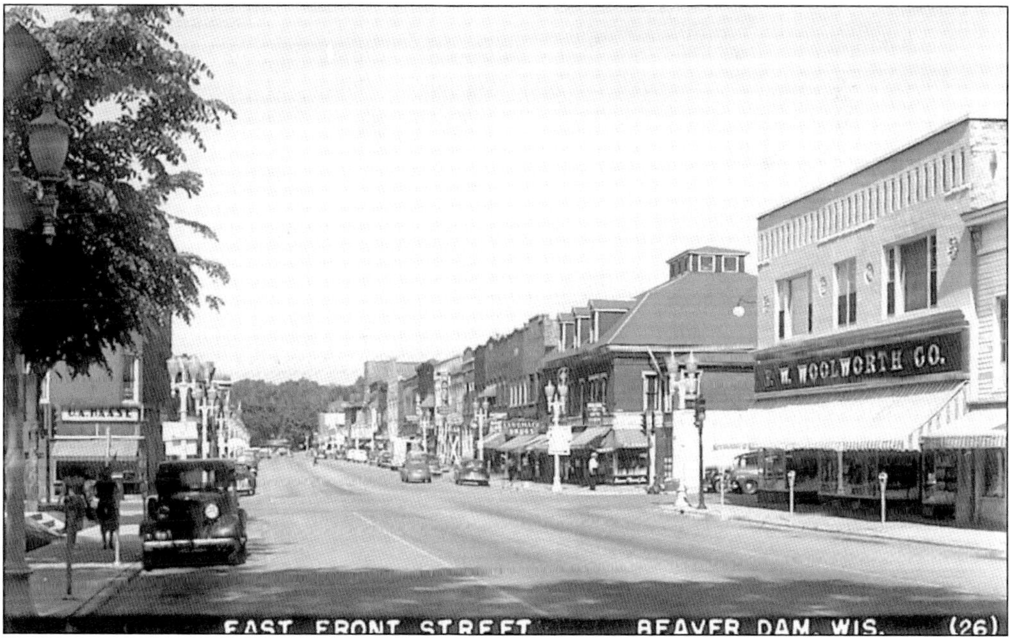

FRONT STREET, LOOKING WEST, 1950S. In the 1950s, the F. W. Woolworth store occupied the northeast corner of the street. Parking meters dotted the sidewalks, and drivers were required to parallel park.

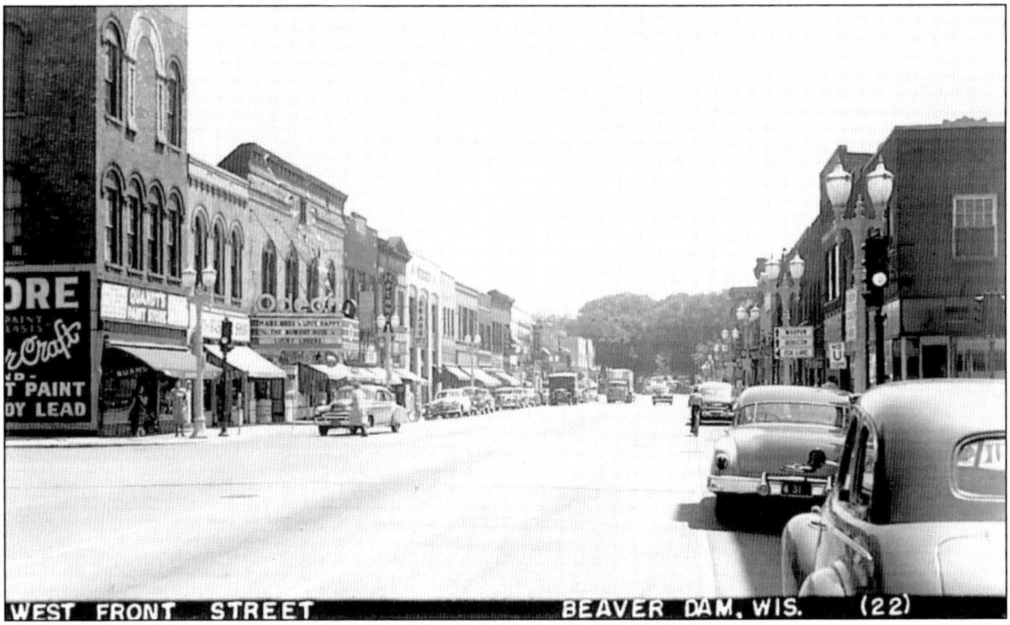

FRONT STREET, LOOKING EAST, 1950S. As the years passed, Quandt's Paint Store replaced Rassman Hardware, and the Odeon evolved from a nickelodeon into a movie palace.

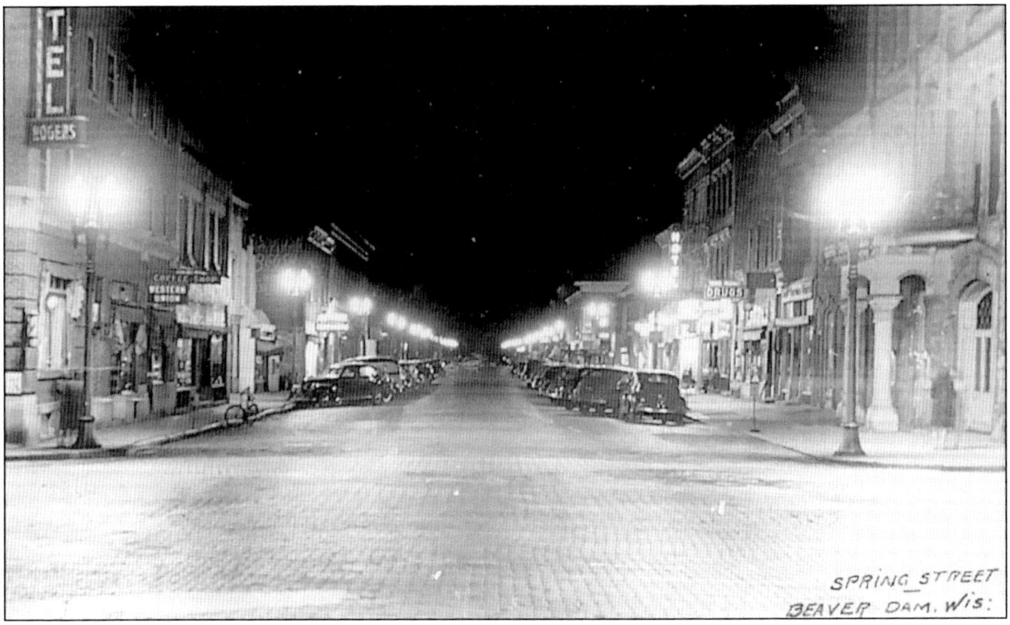

CITY HALL CORNER, LOOKING SOUTH, 1940S. In the 1940s, the corner of Spring and Maple Streets was still dominated by the imposing city hall building, which was complemented by the Hotel Rogers across the street. Unlike Front Street, these streets had not yet been paved with concrete, and the old brick roadways remained.

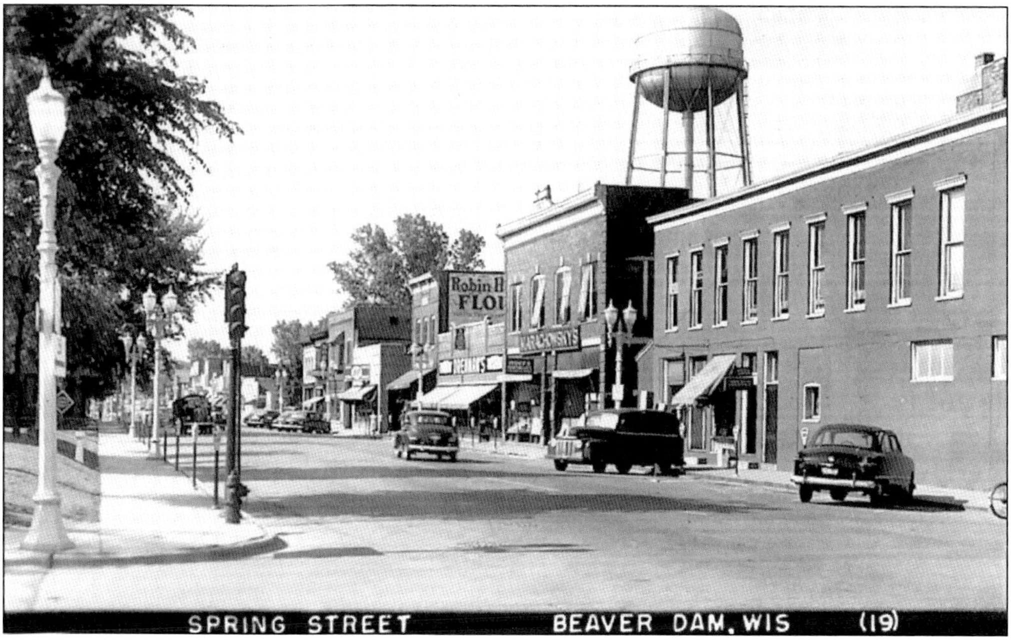

SPRING STREET, LOOKING SOUTH, 1950S. Over the years, the west side of South Spring Street changed dramatically. Wade's Livery was replaced by the Healy Building, which housed the chain grocery store Marchowsky's. Farther down the street were Drennans Department Store, the City Cab Company, the A&P grocery, and the Cleary Motor Company.

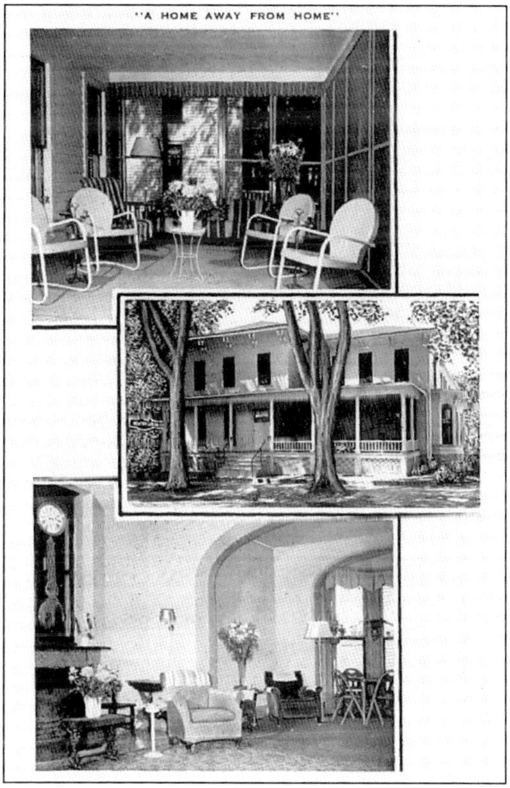

THE NEWTON LODGE. In the early 20th century, this frame residence became the Hotel Newton after an addition provided 20 rooms to rent. In 1917, the property was renamed the Newton Lodge. Lawrence and Laura Heffron purchased the 112 Washington Street building in 1946 and renamed it the Heffron White House.

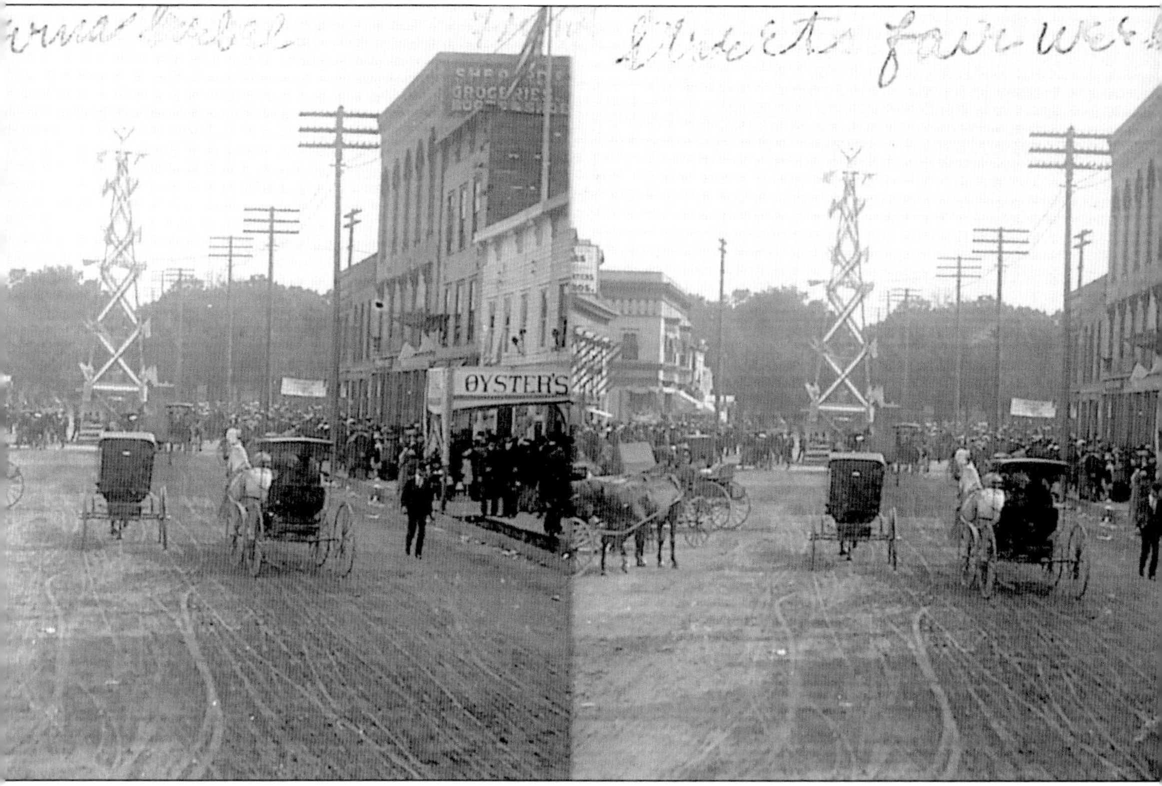

The Oldest Front Street Postcard. Although it was postmarked in 1905, this rare stereoscopic postcard dates from the 1890s. The date of the photograph is evidenced by the dirt roads, the wooden sidewalks, the horse-drawn carriages, and the oyster sign to the right. This photograph was taken during fair week; the wooden structure in the middle of Front Street was erected every year during the fair for a fireworks display.

Three

Summertime and the Livin' Is Easy

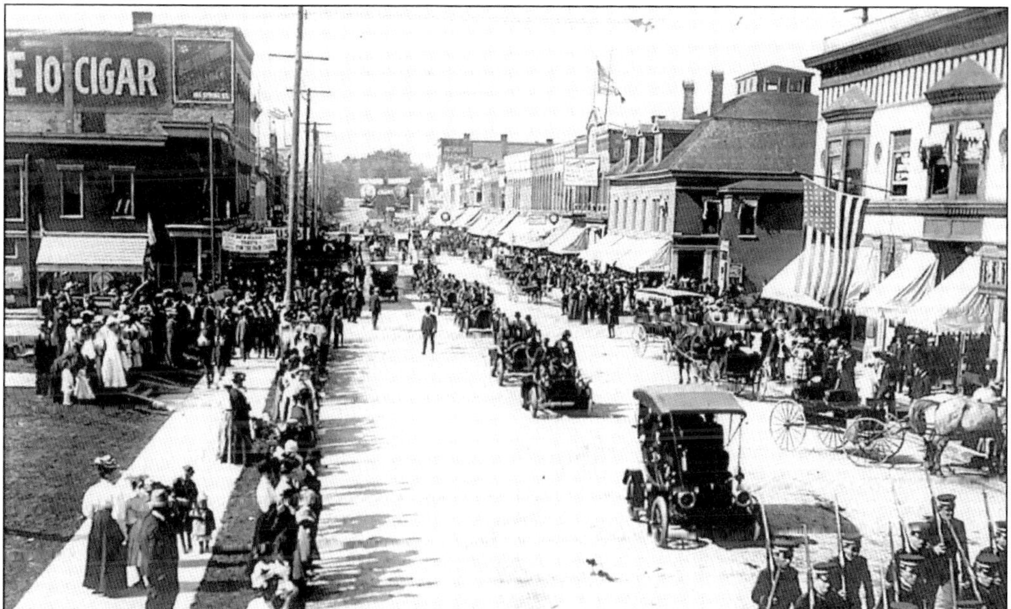

July 4, 1905. The Fourth of July was always a special day of celebration for Beaver Dam residents. Activities included a parade, a citywide picnic, and horse racing at the fairgrounds.

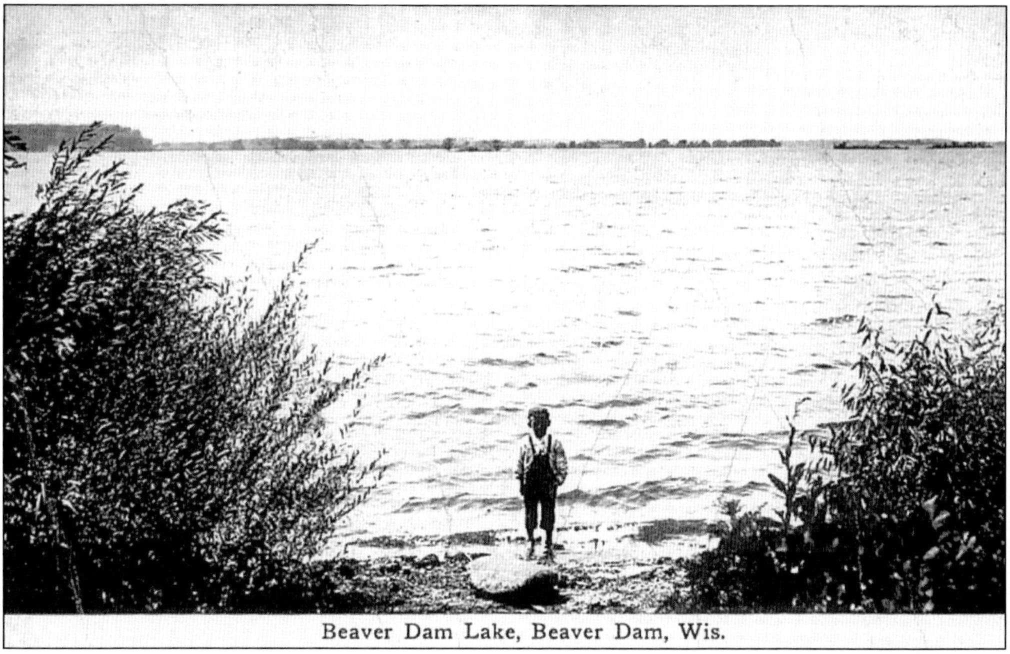

BEAVER DAM LAKE. The man-made Beaver Dam Lake has played a pivotal role in the city's industry, recreation, and politics since its creation in 1843.

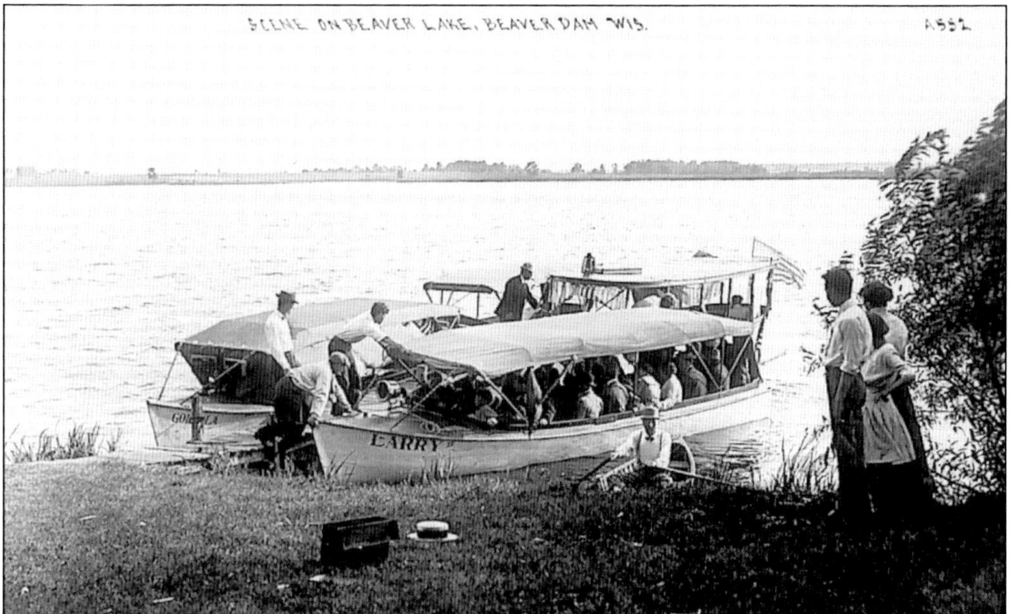

THE *LARRY*. At the turn of the century, Beaver Dam Lake boasted several steamboat excursion lines. The first steamboat was the *Belle Rowell.* Others included the *J. S. Rowell* and the *Rambler.* The *Larry* was a part of the second generation of gas-powered excursion ships.

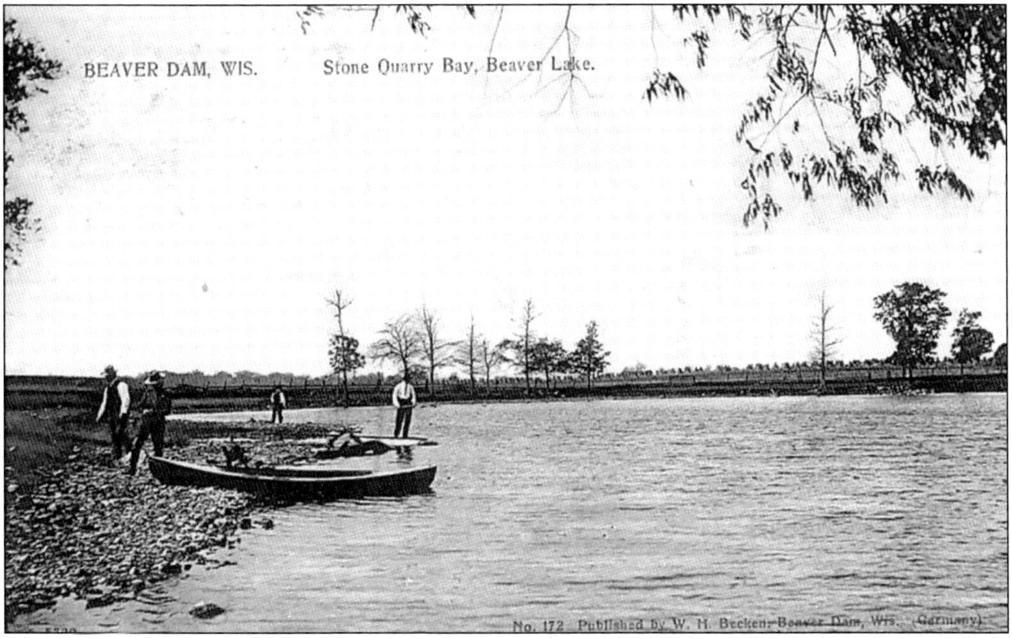

STONE QUARRY POINT. Stone Quarry Point was a popular place for anglers. It was located on the west side of Beaver Dam Lake near County Highway CP.

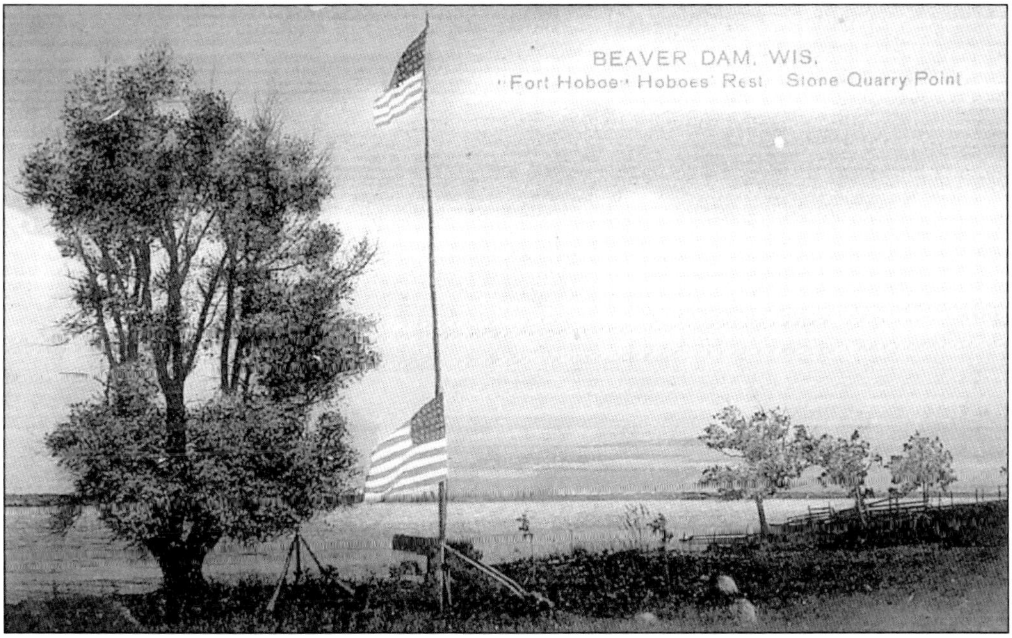

FORT HOBOE. Fort Hoboe, a tribute to veterans of the Civil and Spanish-American Wars, was established at Stone Quarry Point.

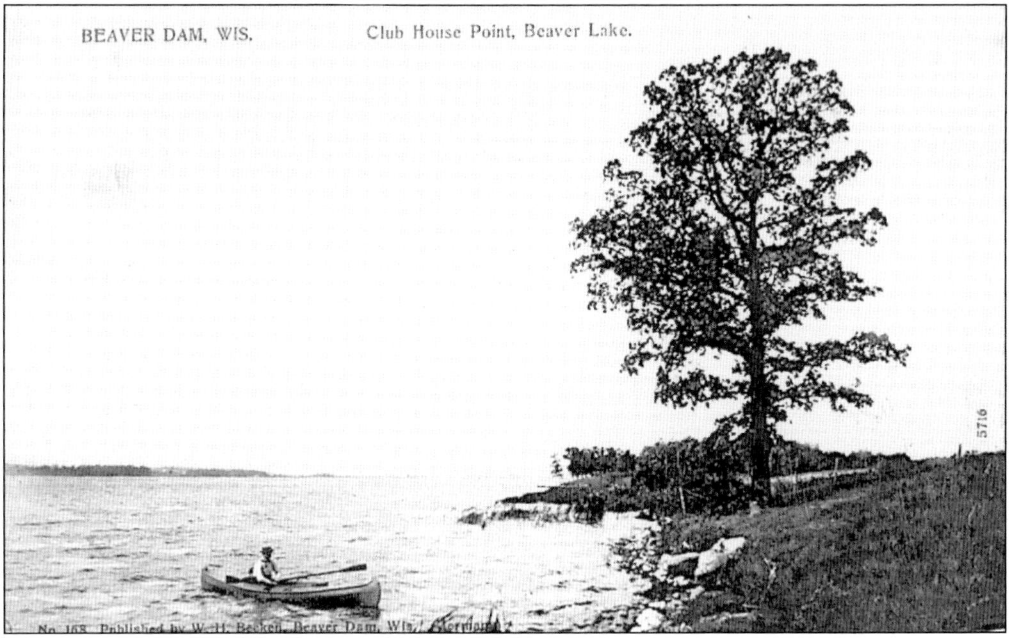

CLUB HOUSE POINT.

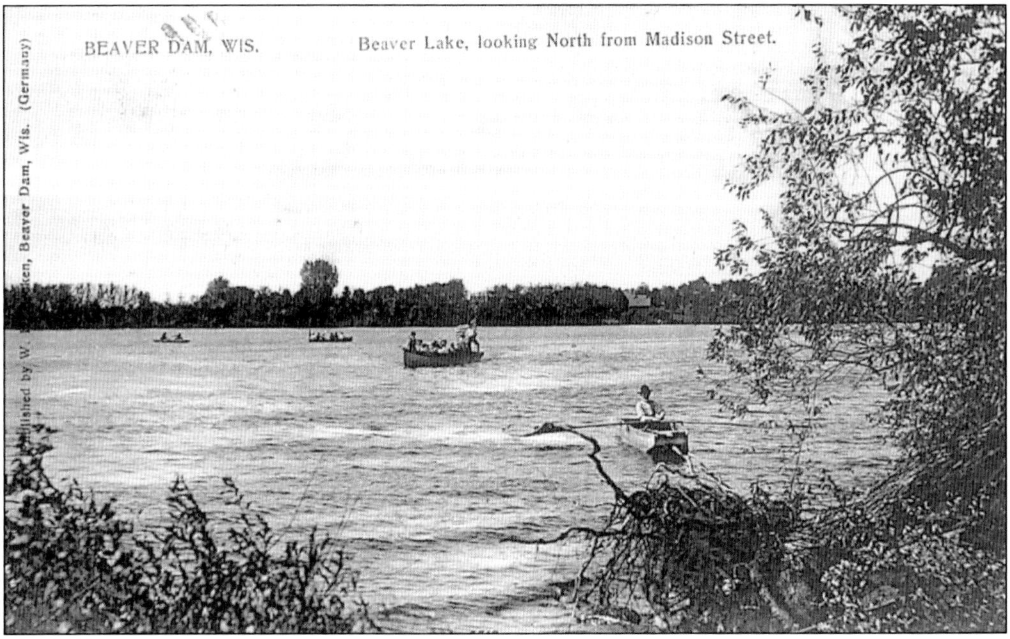

BEAVER LAKE FROM MADISON STREET. Beaver Dam Lake was originally called simply Beaver Lake. It is unclear when the name change took place.

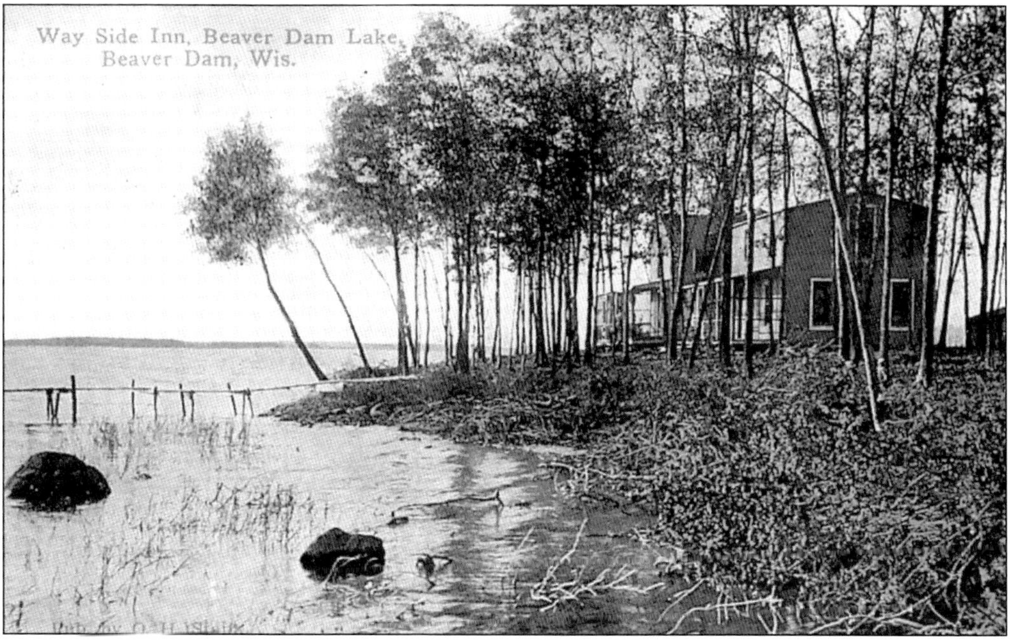

THE WAYSIDE INN. The Wayside Inn welcomed tourists searching for the finest fishing holes and hunting grounds. The inn was located on Beaver Dam Lake northeast of Goose Island.

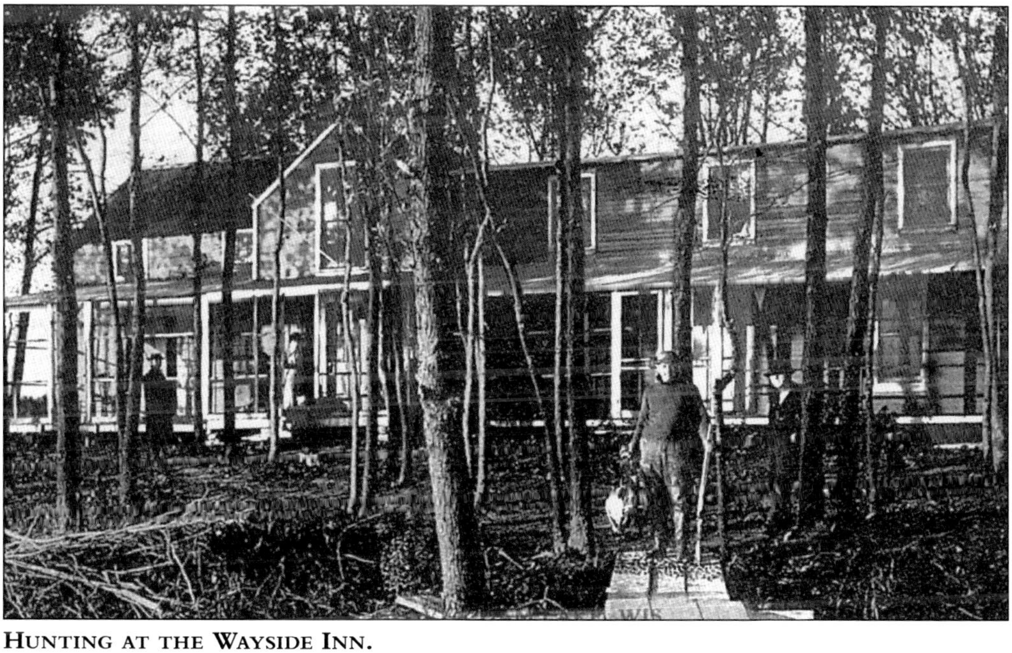

HUNTING AT THE WAYSIDE INN.

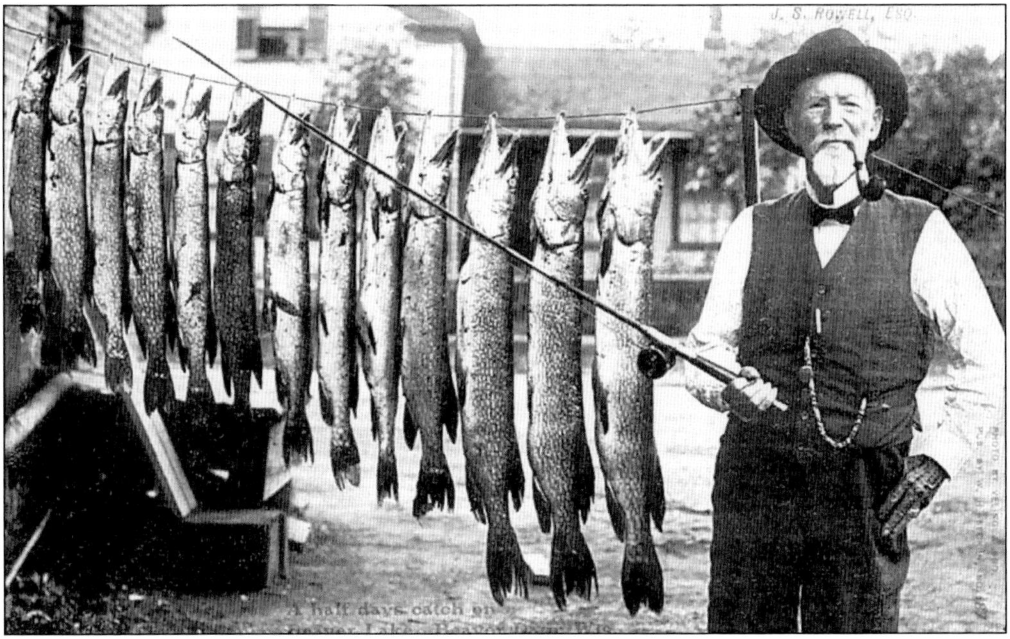

FISHERMAN J. S. ROWELL. One of the great industrialists of Beaver Dam, J. S. Rowell was also an accomplished angler. Rowell founded the J. S. Rowell Manufacturing Company, an agriculture implement firm, in 1855. He created the famous "Tiger Seeder." The business became one of the city's great industrial success stories, lasting 57 years.

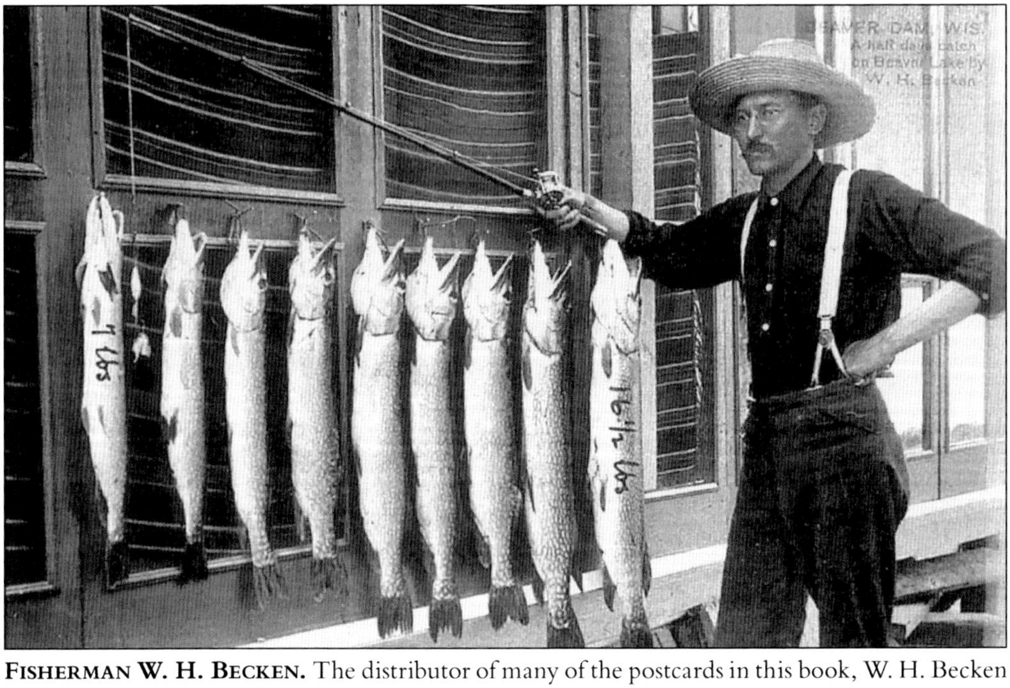

FISHERMAN W. H. BECKEN. The distributor of many of the postcards in this book, W. H. Becken could also handle a rod and reel with great finesse. Becken is also featured on page 26.

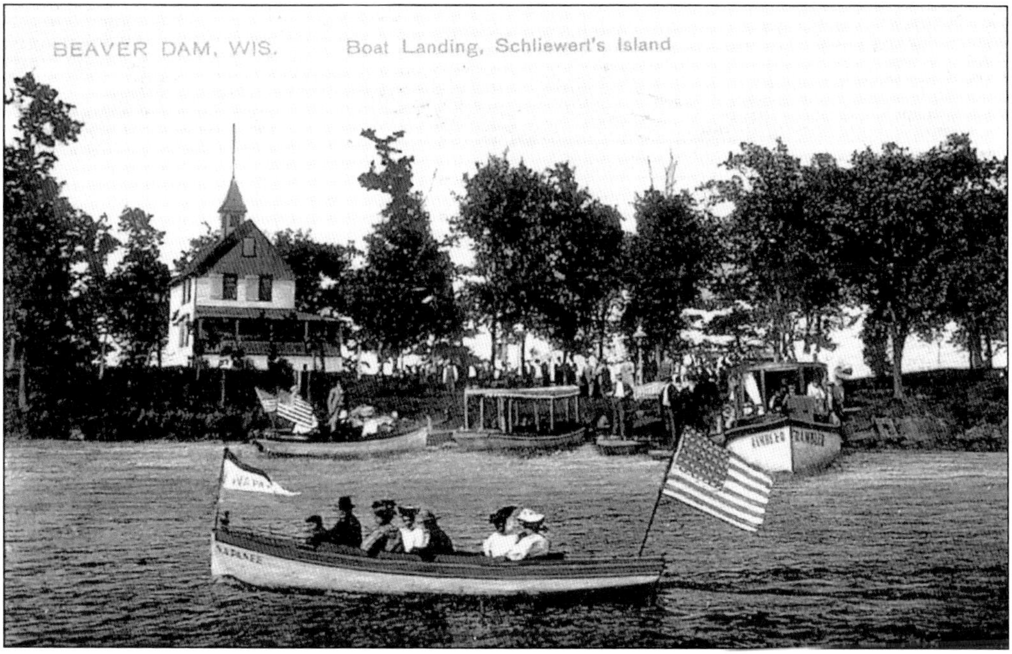

SCHLIEWERT'S ISLAND. Another favorite recreation spot on the lake was Schliewert's Island Resort. The island was located along the northeast shore of Beaver Dam Lake.

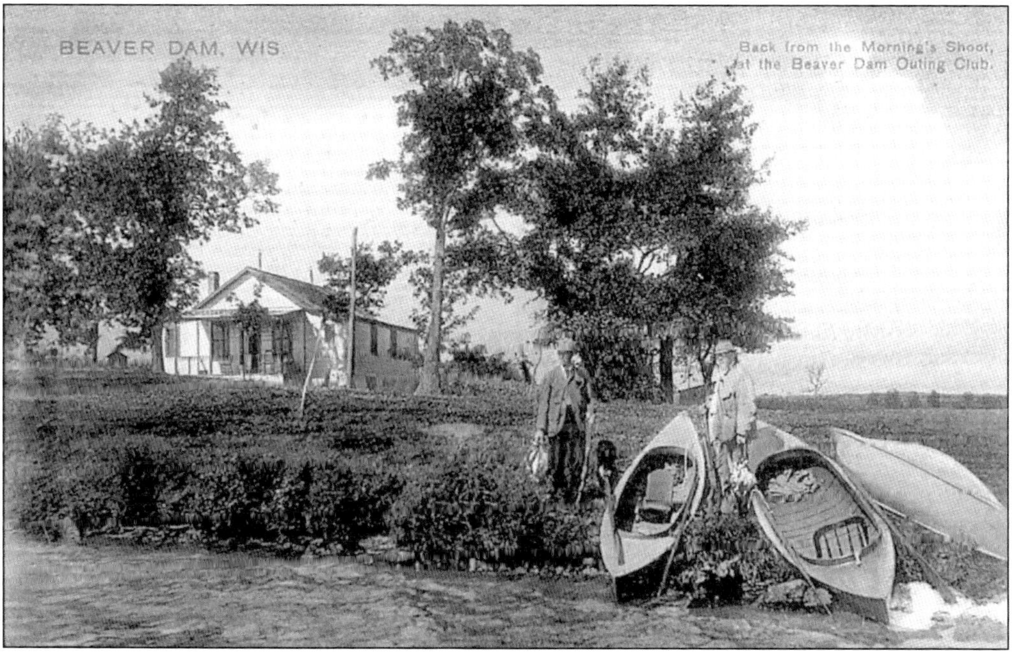

THE OUTING CLUB. Local gentlemen frequented the Outing Club to bag their limit of game birds. Note that these huntsmen's morning attire consists of suits, white shirts, ties, and hats.

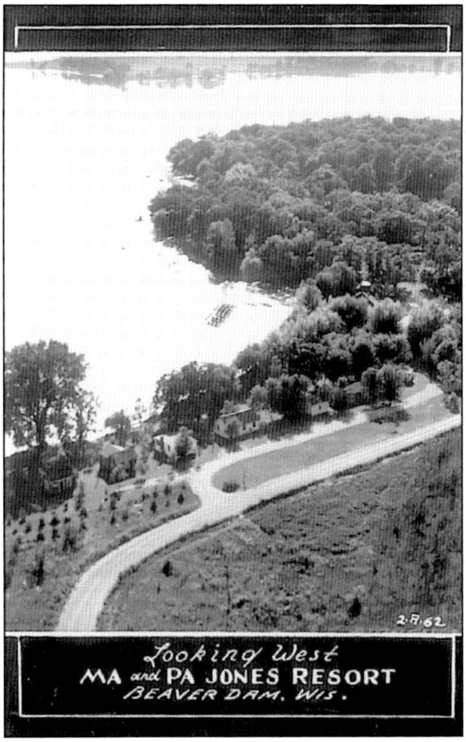

THE MA AND PA JONES RESORT. One of the longest operating resorts on Beaver Dam Lake was the Ma and Pa Jones Resort, located off Highway 33 at Sunset Point. Its slogan was "Pa's a crab; Ma's a honey."

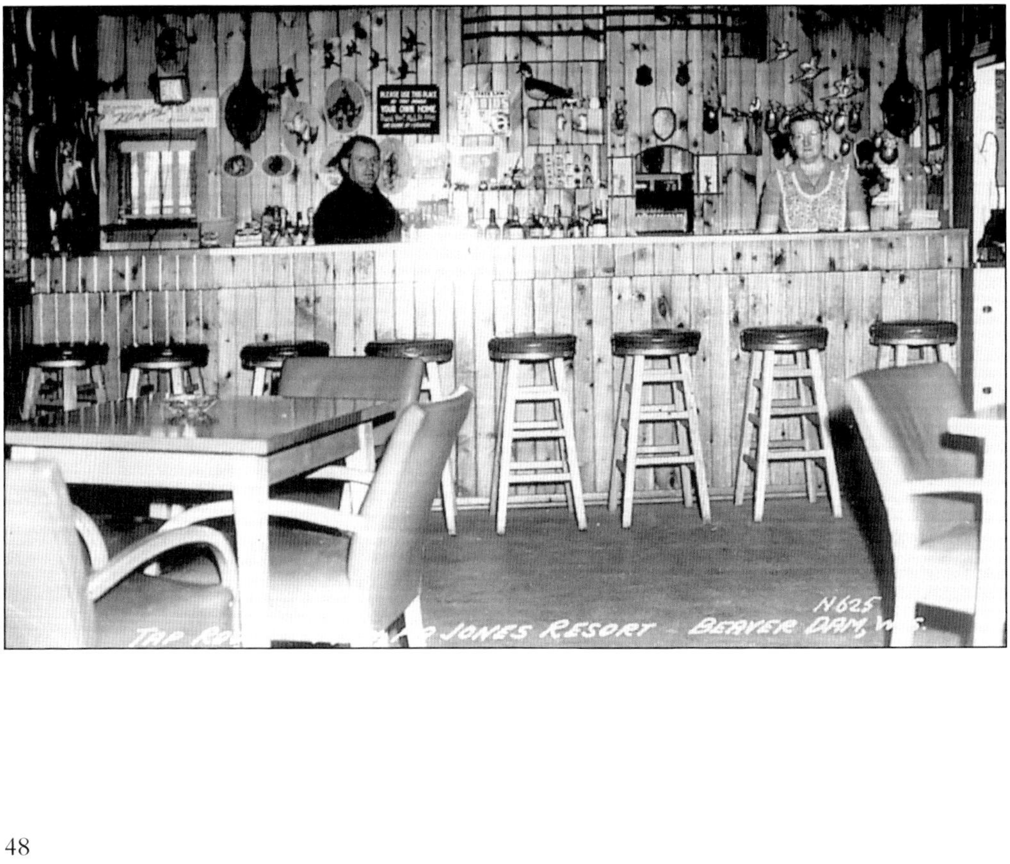

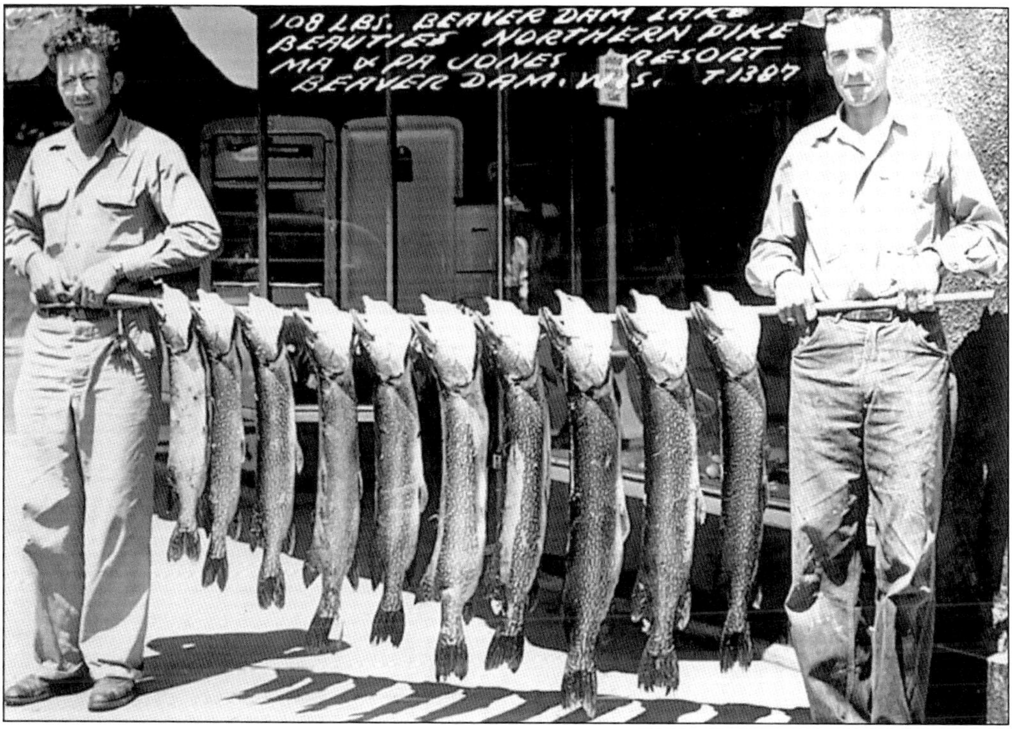

A DAY'S CATCH. Along with the great fishing offered on Beaver Dam Lake, visitors to the Ma and Pa Jones Resort could enjoy bullhead skinning exhibitions by champion skinner Griff Jones, who routinely skinned 10 to 13 bullheads per minute.

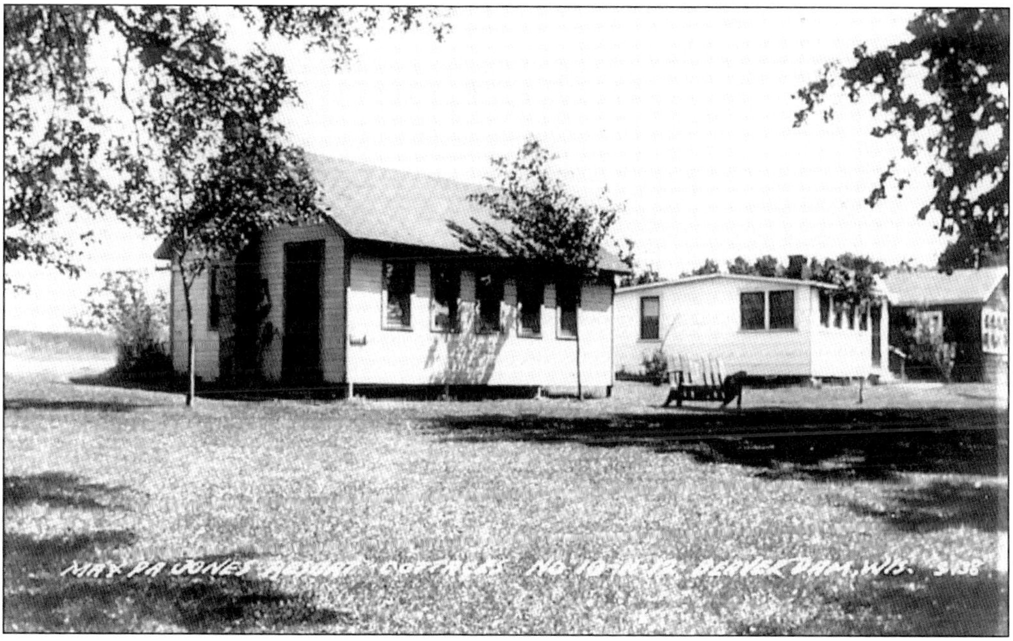

THE MA AND PA JONES RESORT COTTAGES. The resort consisted of a main lodge and 12 cabins and was open year-round.

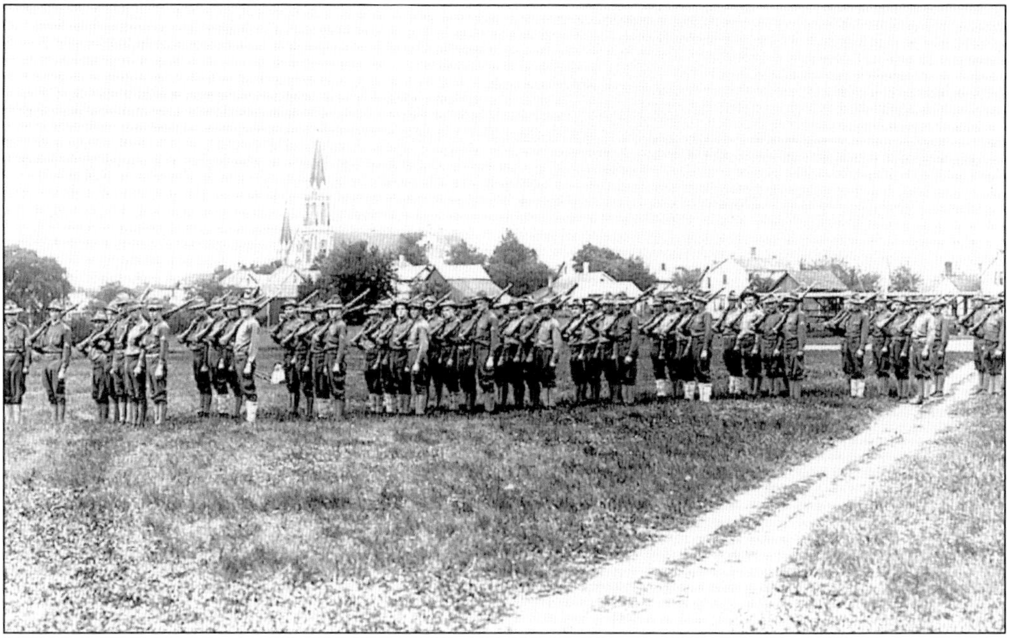

Company K Summer Drills. Company K was originally organized in 1880 as the Burchard Guards. When it mustered into federal service in 1898, the unit became Company K of the 2nd Wisconsin Voluntary Infantry. Company K served valiantly in the Spanish-American War and in World War I.

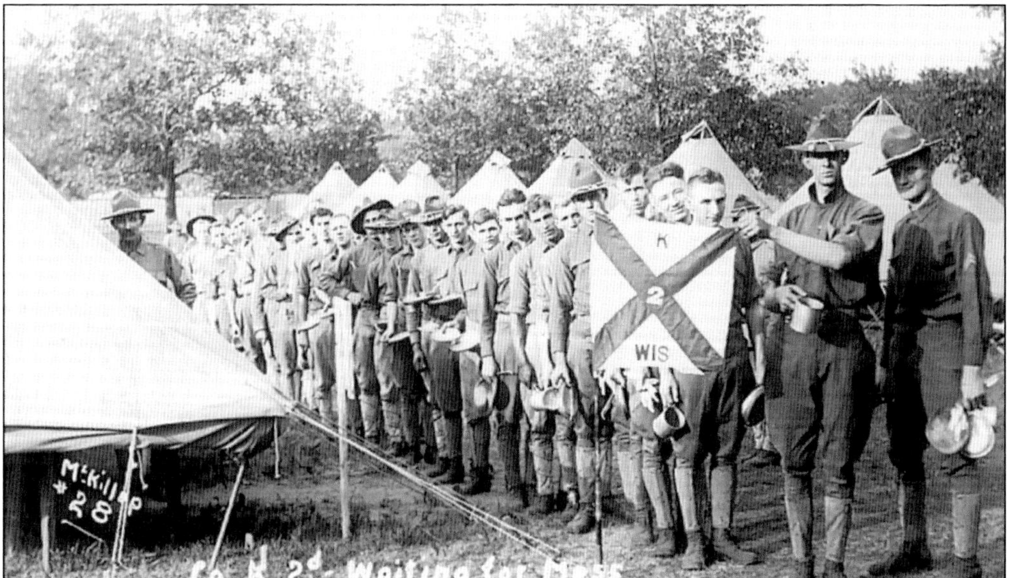

Company K at Camp Douglas. Company K often mustered for maneuvers at Camp Douglas, located near Tomah. After World War I, Company K was reorganized into Company E, 128th Infantry, 32nd Division.

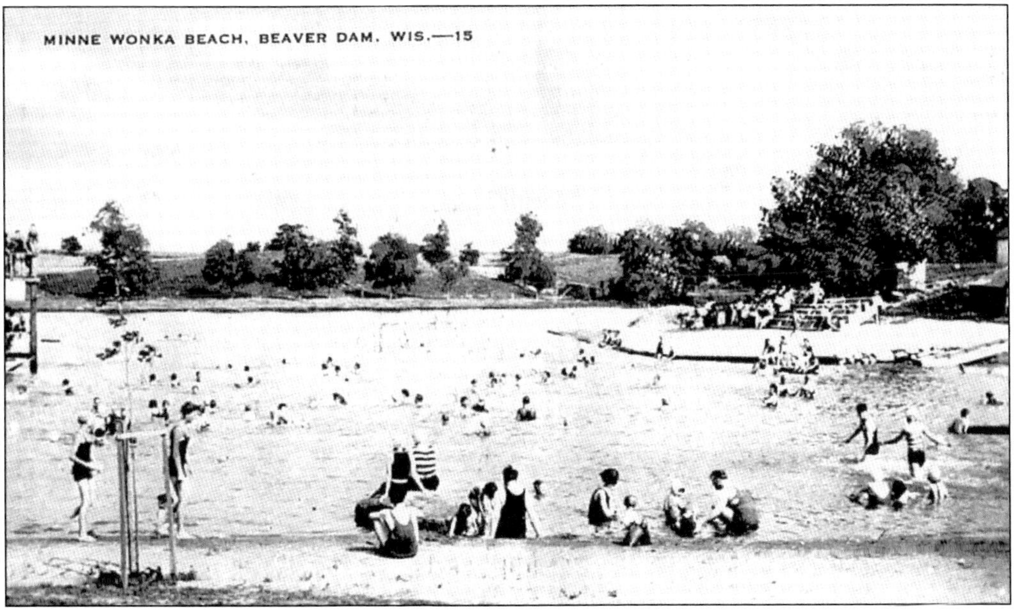

MINNE WONKA BEACH, 1940S. In 1930, the Kiwanis Club purchased 18 acres surrounding Crystal Lake to establish a bathing beach. The area was originally part of the Dr. George Swan Resort. A children's contest was held to choose the new name of the beach, and the winning entry was Minne Wonka, a Native American phrase meaning "place of happiness and contentment."

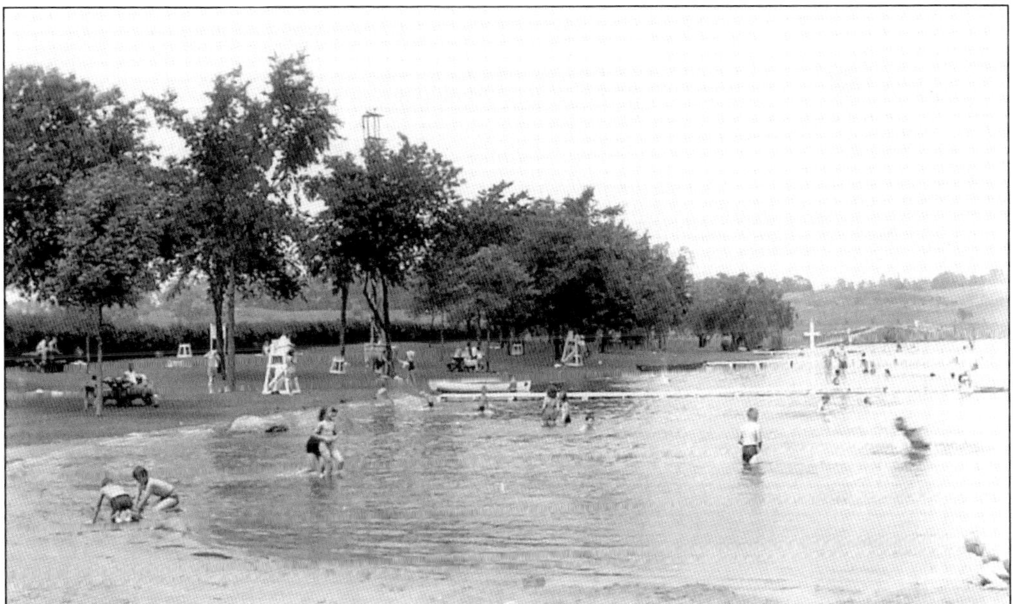

CRYSTAL LAKE BEACH, 1950S. In 1942, the Kiwanis Club donated the beach to the city. Thereafter, the beach reverted to its earlier title of Crystal Lake Beach.

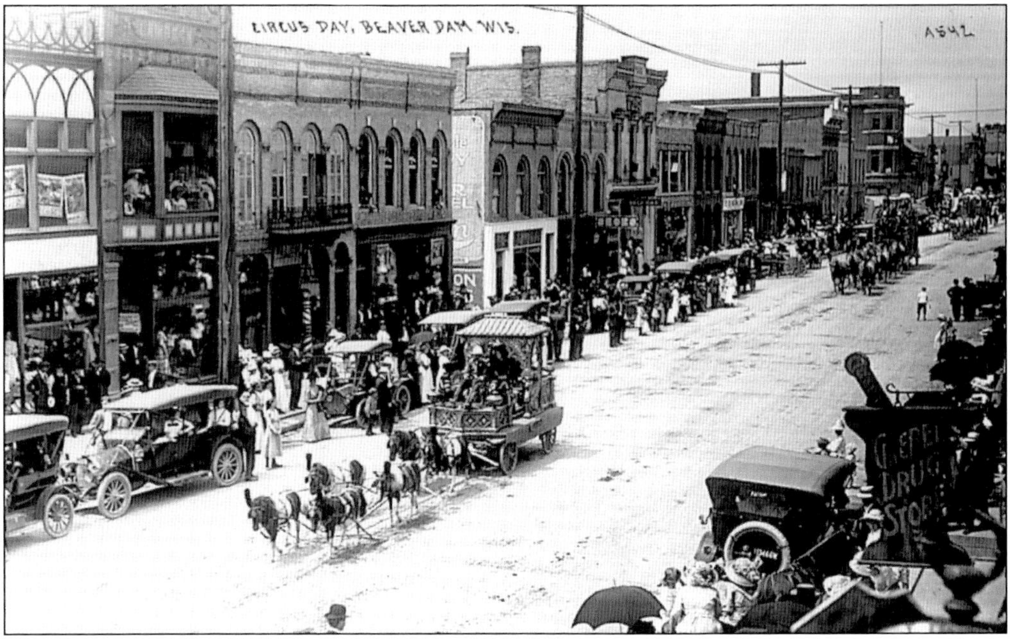

THE CIRCUS COMES TO TOWN. One of the main events of summer in the city was the arrival of the circus. In the mid- to late 19th century, any number of circus troupes, from Wild West shows to touring European big tops, entertained Beaver Dam audiences of all ages.

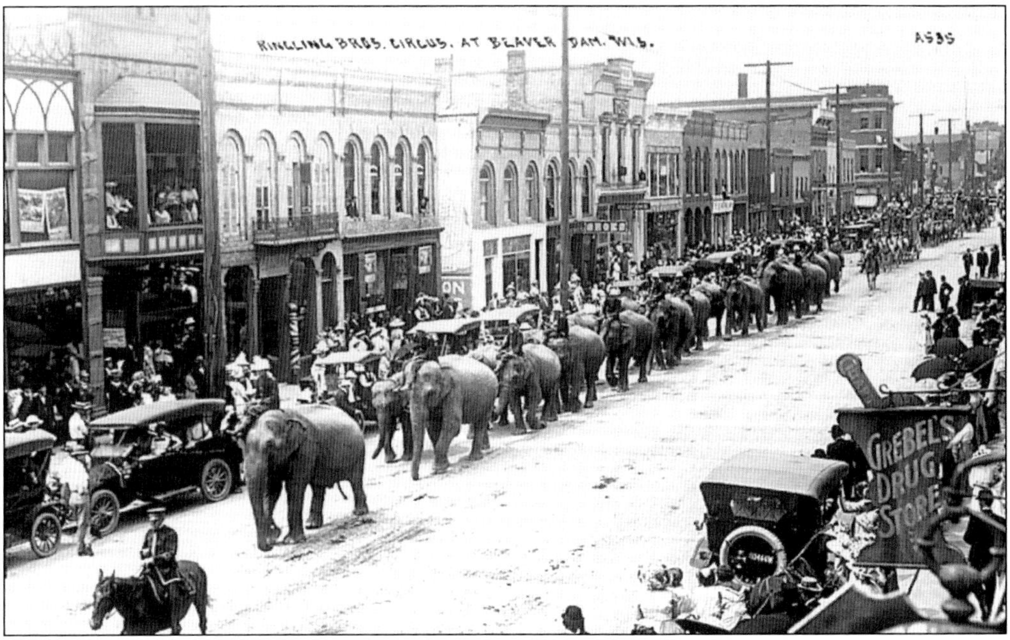

THE RINGLING BROTHERS CIRCUS. By the dawn of the 20th century, the major troupes, traveling by rail, brought amusement to the city during the summer.

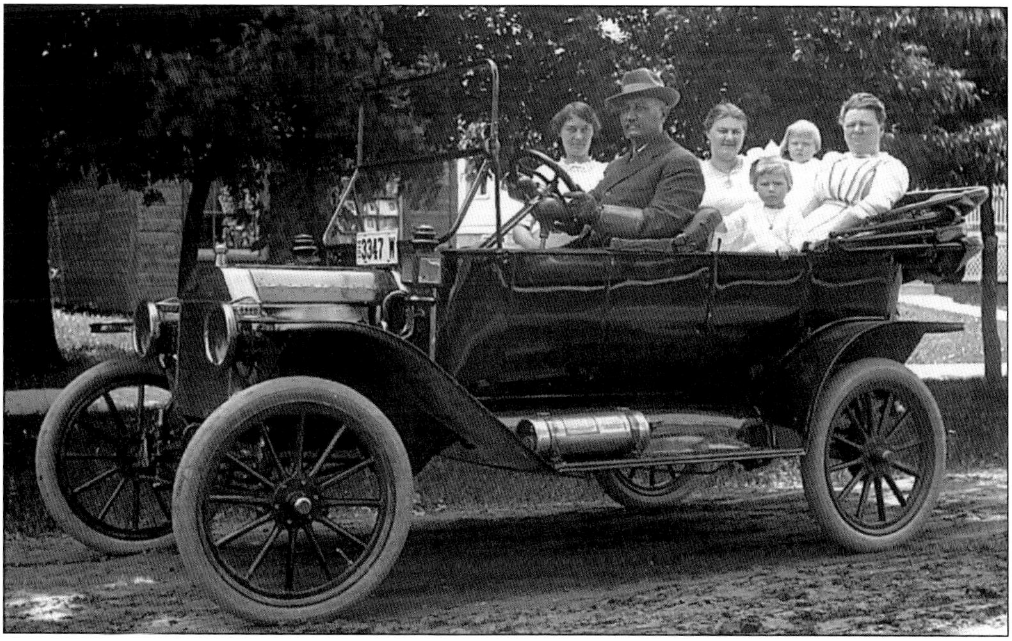

A SUNDAY DRIVE. After their homes and families, people's pride and joy were their automobiles. In the early 20th century, a Sunday afternoon was the perfect time to impress the neighbors and take a tour of the city.

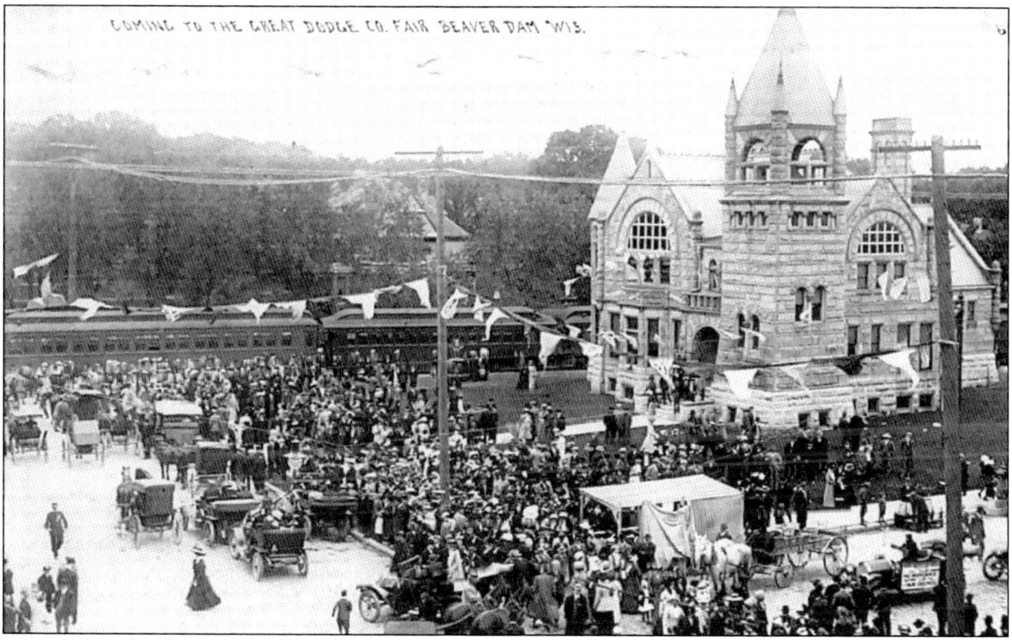

COMING TO THE FAIR. The Dodge County Fair was one of the most popular attractions of the late summer and early fall. In 1902, some 70,000 people attended the fair. Special trains were commissioned to transport fairgoers from Chicago and Milwaukee.

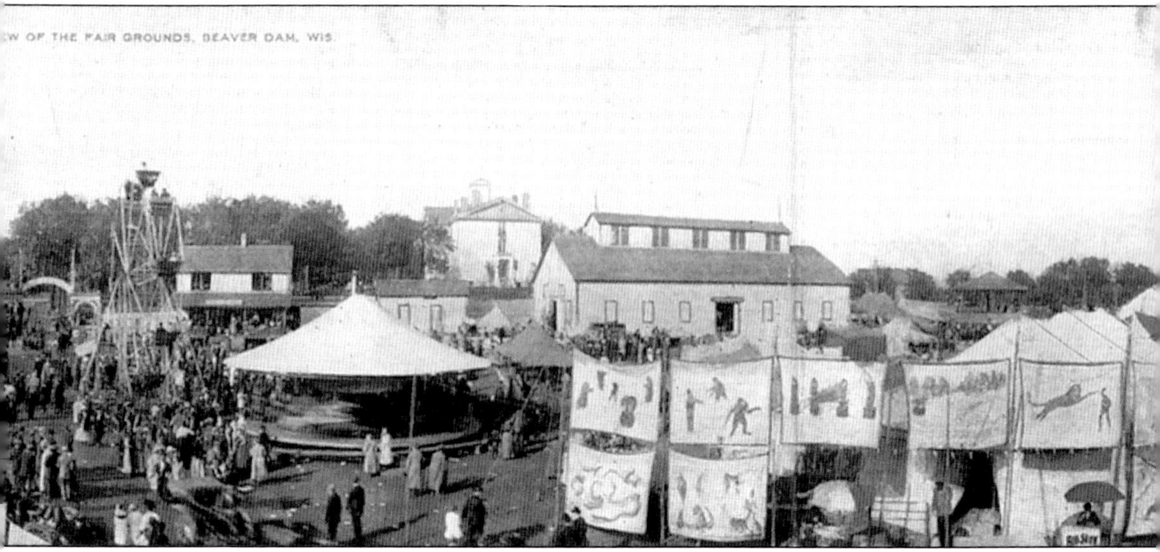

THE WORLD'S GREATEST COUNTY FAIR. Previously held in Juneau, the Dodge County Fair found a permanent home in Beaver Dam in 1886. Located at the corner of Park and University

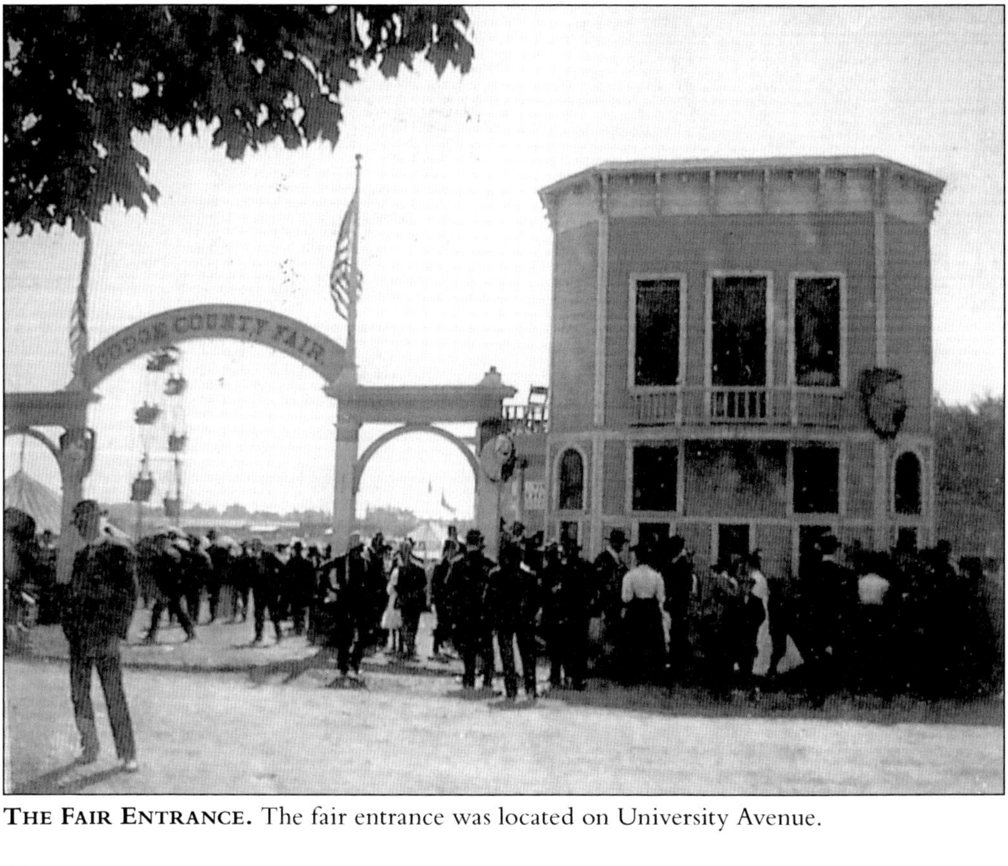

THE FAIR ENTRANCE. The fair entrance was located on University Avenue.

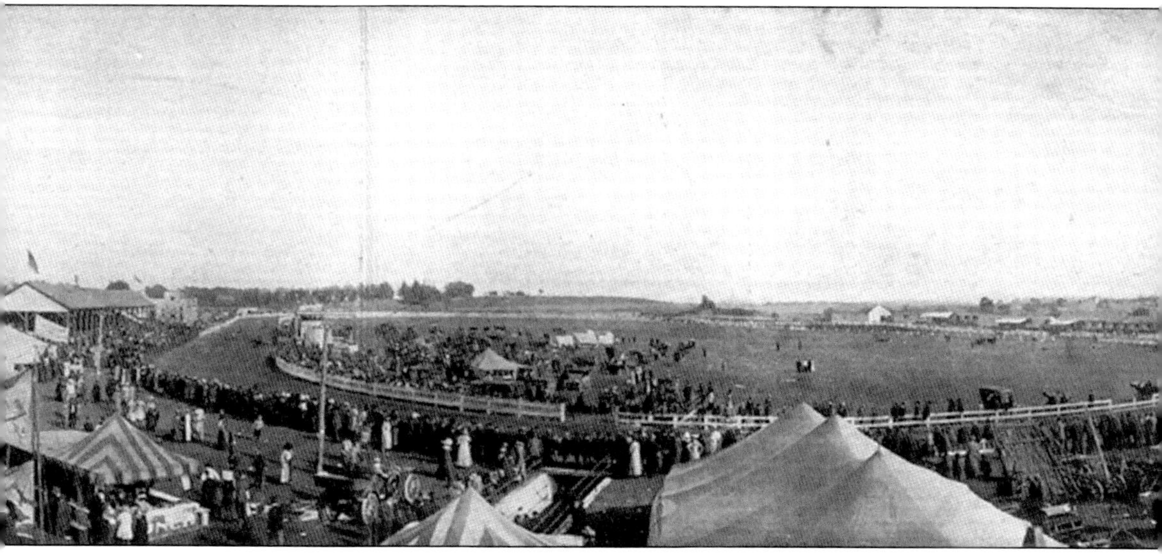

Avenues, the weeklong affair offered the finest in entertainment and adopted the moniker "World's Greatest County Fair."

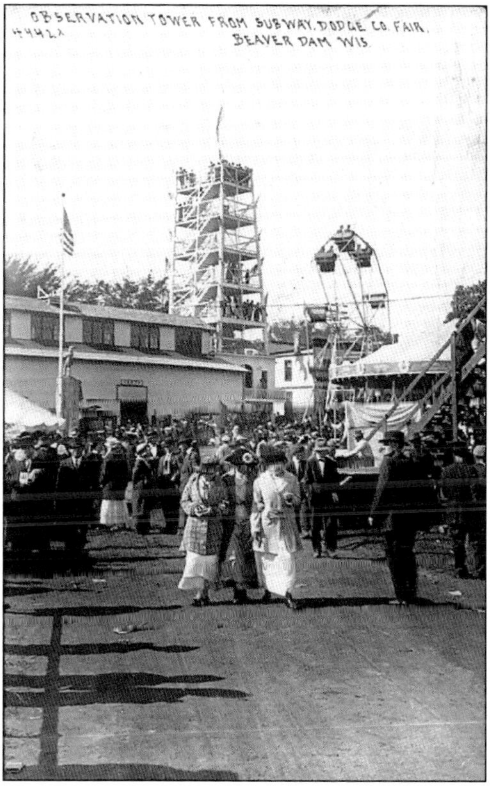

THE OBSERVATION TOWER. An observation tower was located at the fair's entrance to allow fair officials and the occasional lucky patron a complete view of the fair activities.

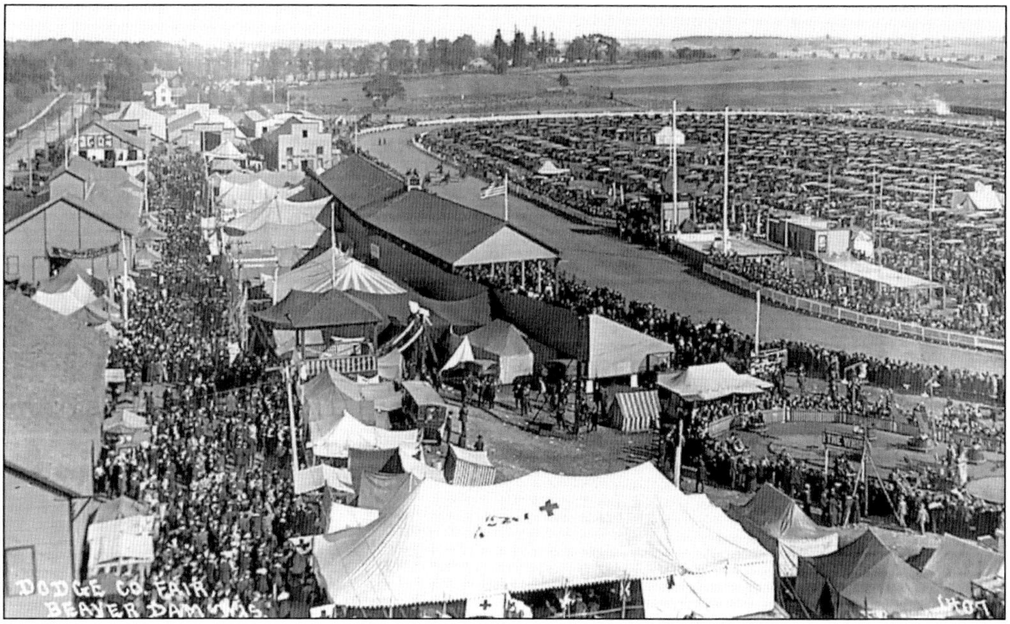

THE MIDWAY AND THE FERRIS WHEEL. The World's Greatest County Fair secured the finest in midway attractions.

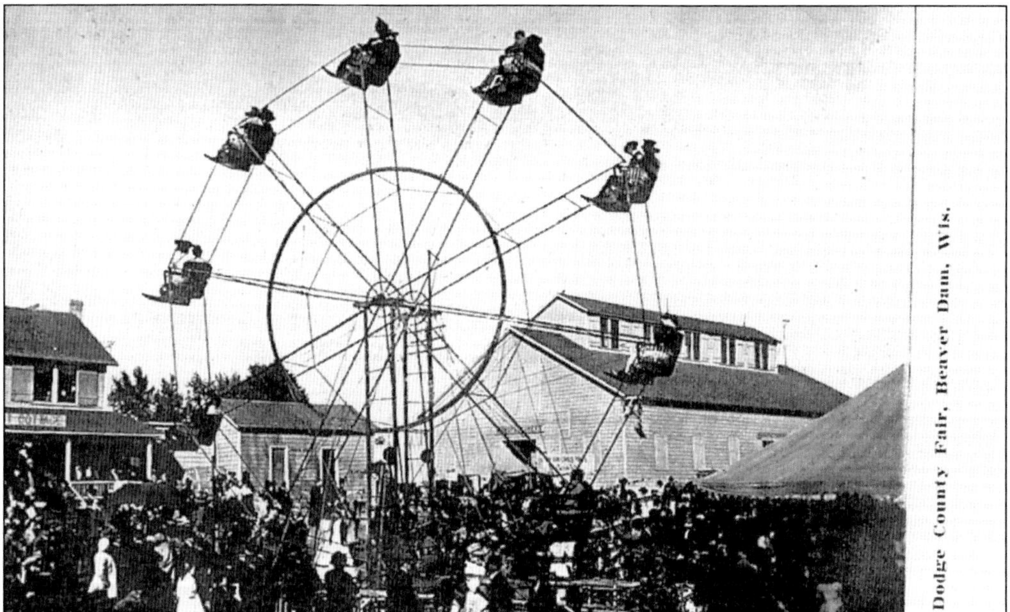

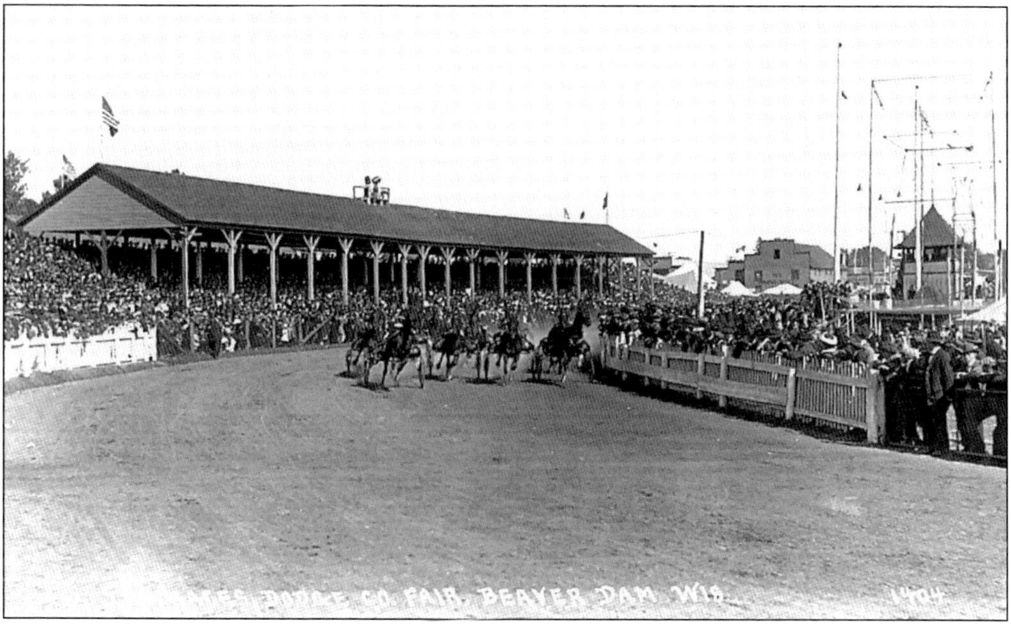

HARNESS RACING. When the fairgrounds were purchased by the Dodge County Agriculture Society, the deed contained a restriction that no horse racing take place on the premises during the grantor's lifetime. The society assured the grantor that the track that was to be built on the property was a location for driving their horses, not for racing. Soon thereafter, the grantor met his maker, and the track was immediately utilized for harness competitions.

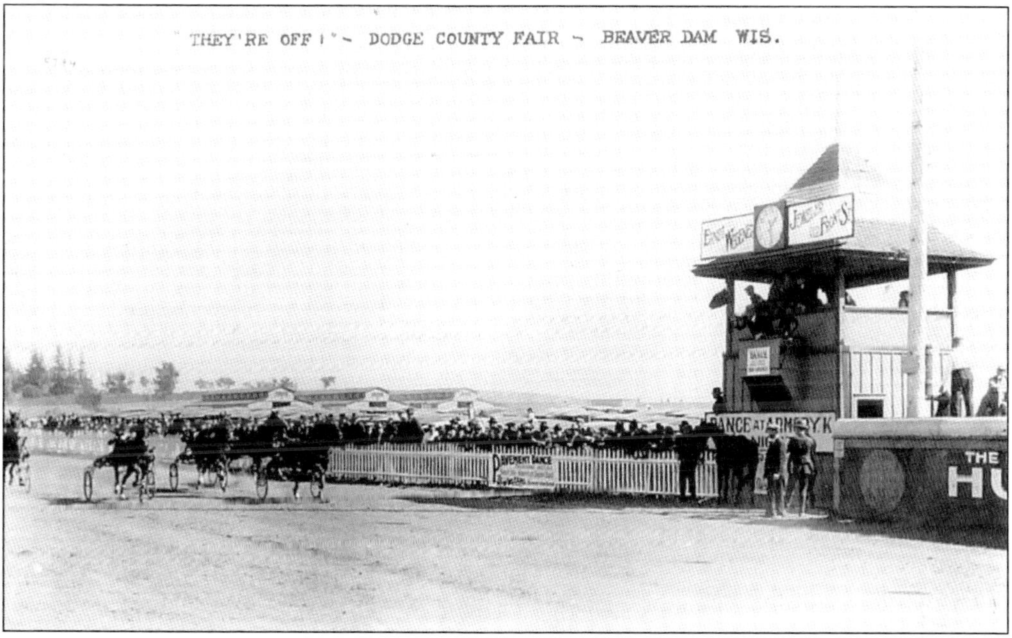

THE TRACK TOWER. The fair went to great lengths to employ the services of the Midwest's elite announcers. The race was called from this tower.

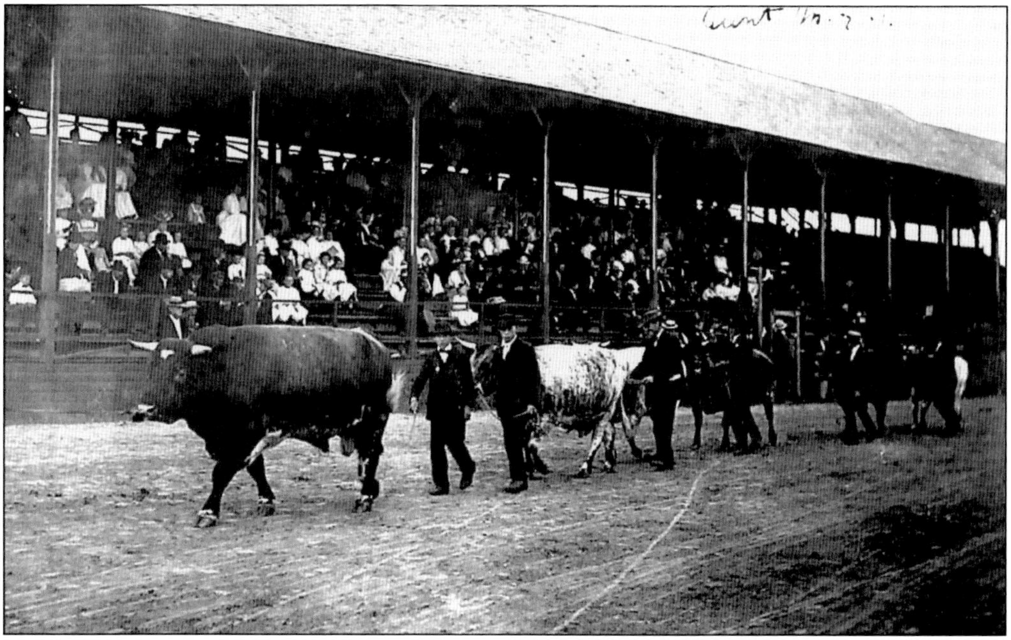

A Livestock Competition. The Dodge County Fair has long been recognized as a premier venue for the exhibition of the finest livestock in the state.

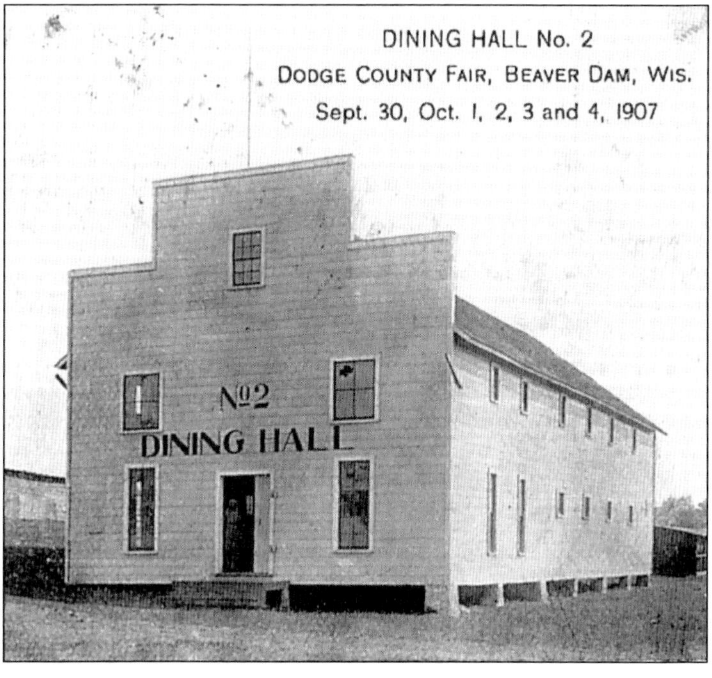

The Dining Hall. The fairgrounds consisted of a complex of buildings that provided housing for livestock, exhibit space for local tradesmen, and dining facilities for the multitude of fairgoers. These large dining halls were used throughout the year to host large gatherings. For example, Wayland held its semicentennial dinner in the fairground dining hall in 1905.

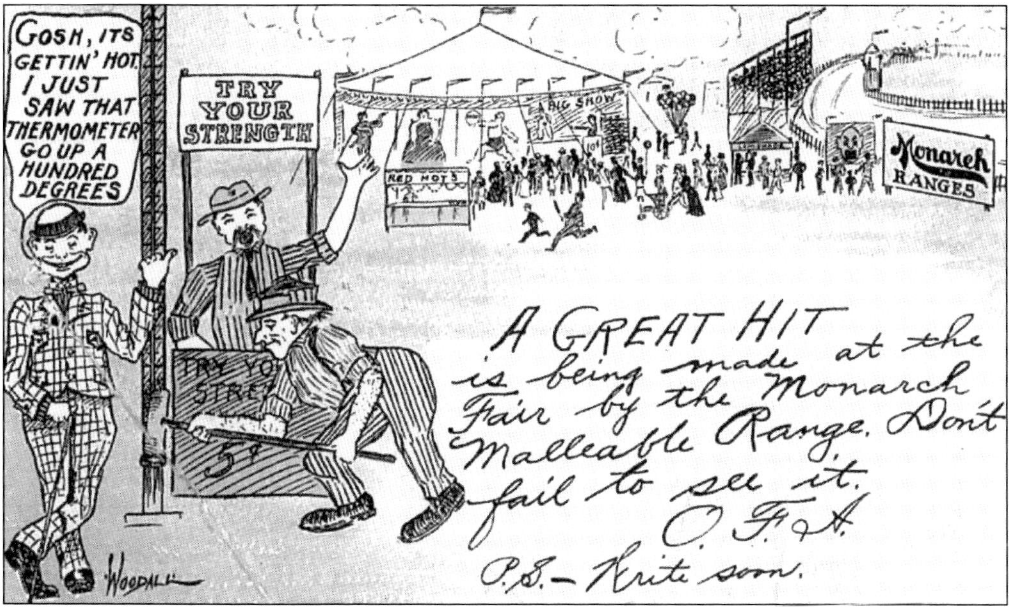

MONARCH PROMOTIONAL POSTCARDS. These two postcards invited potential fairgoers to visit one of the many merchants exhibiting during fair week.

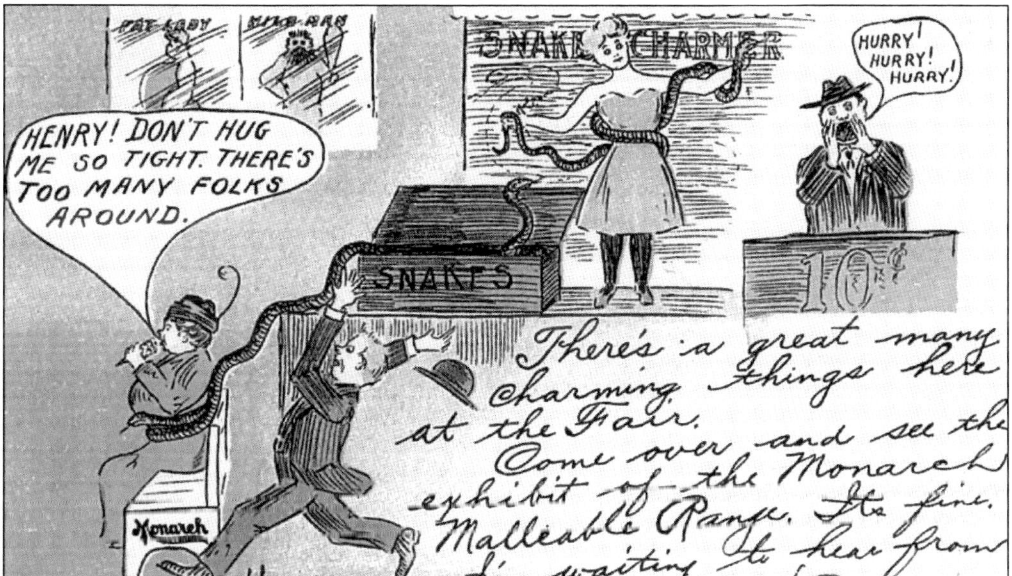

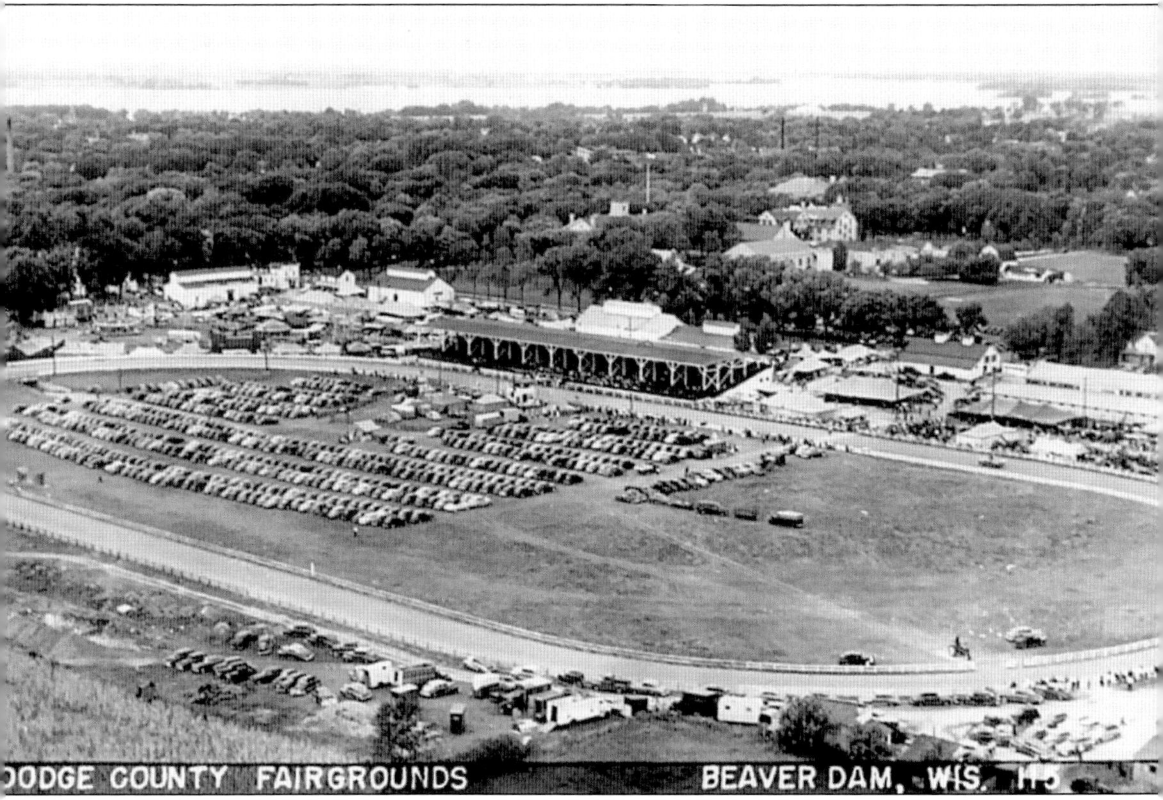

Nearing the End, 1950s. The fair's run at the University Avenue site was nearing its end in the late 1950s. Due to the limited space, fair officials began contemplating alternative locations. The decision was expedited when, in 1958, a substantial portion of the wooden stands pictured above were destroyed by fire. The fair remained at the location for two more years, moving to its current site in 1960.

Four

On the Street Where You Live

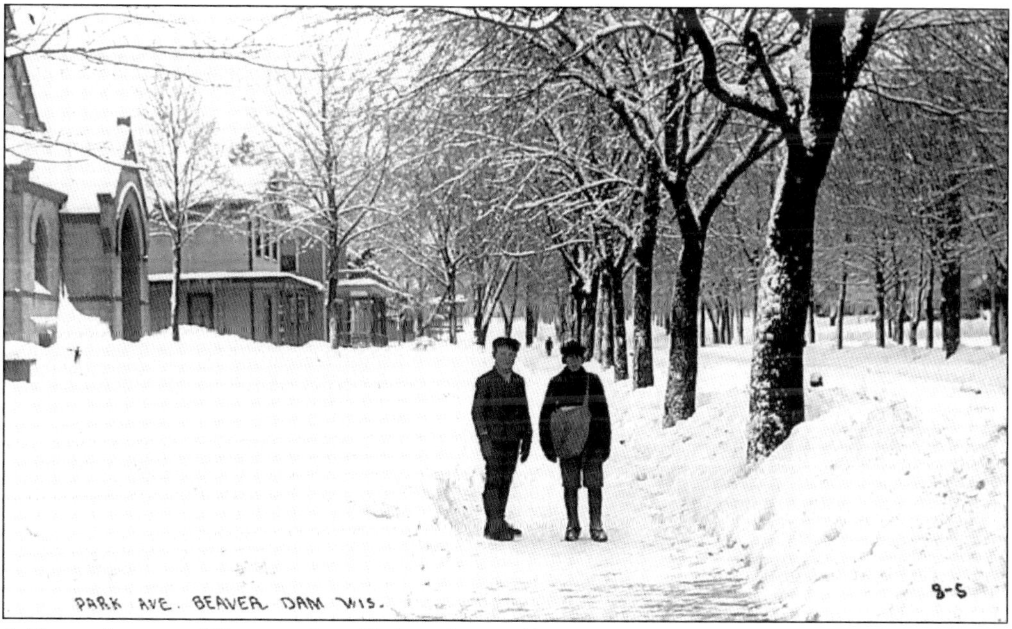

A Park Avenue Winter. Bundled up for a cold winter day, two boys deliver newspapers on Park Avenue in 1910. The church to the left is the Grace Presbyterian Church.

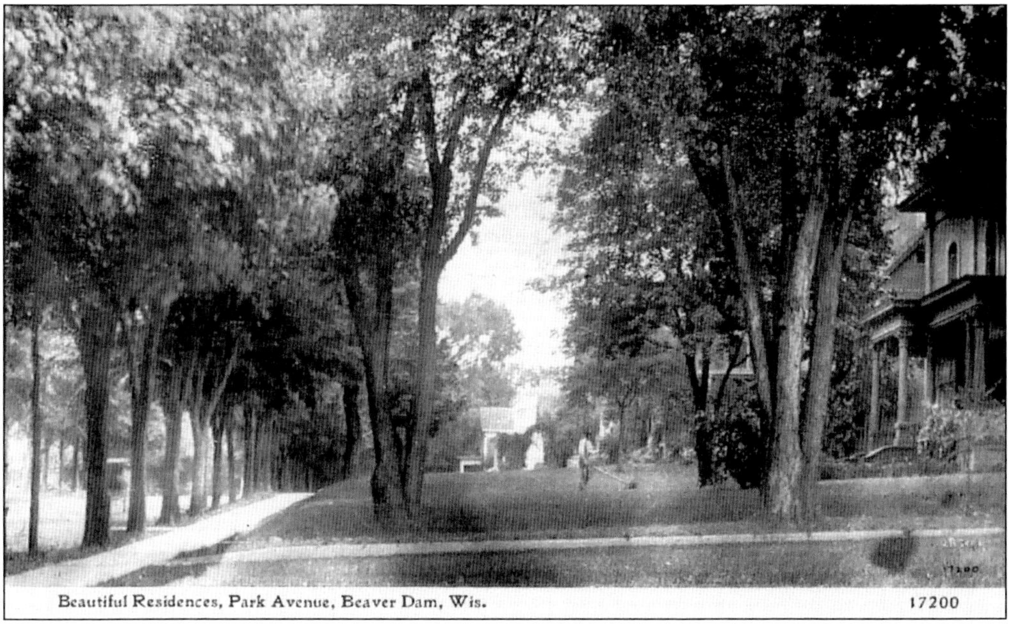

A Park Avenue Summer. This photograph was taken from the corner of Park and Lincoln Avenues, looking east along the south side of Park Avenue.

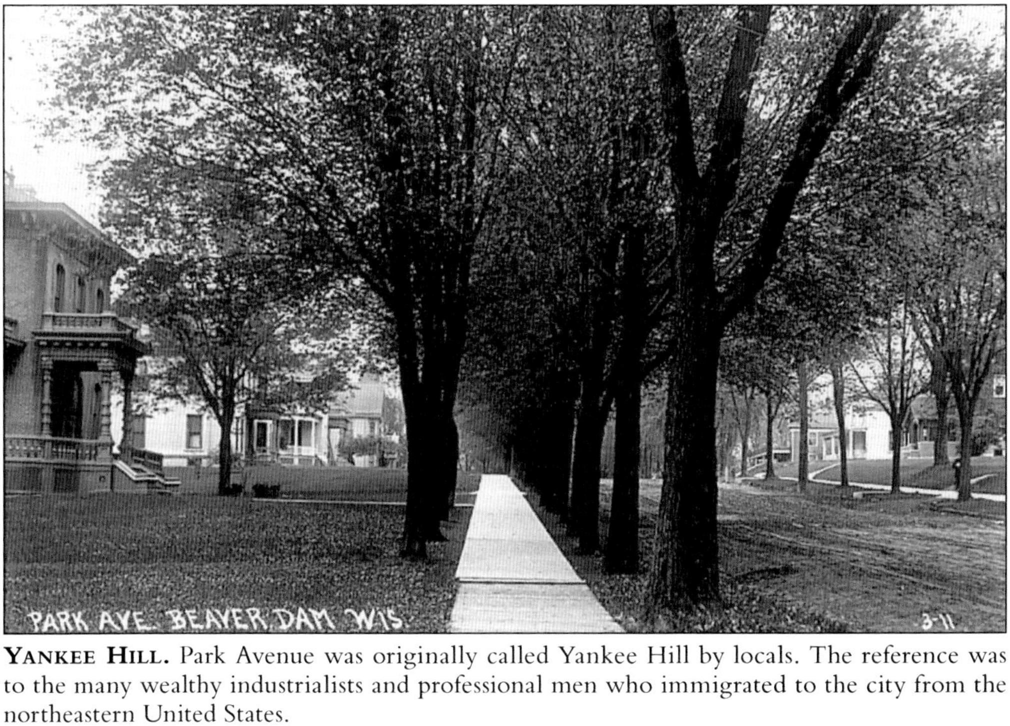

Yankee Hill. Park Avenue was originally called Yankee Hill by locals. The reference was to the many wealthy industrialists and professional men who immigrated to the city from the northeastern United States.

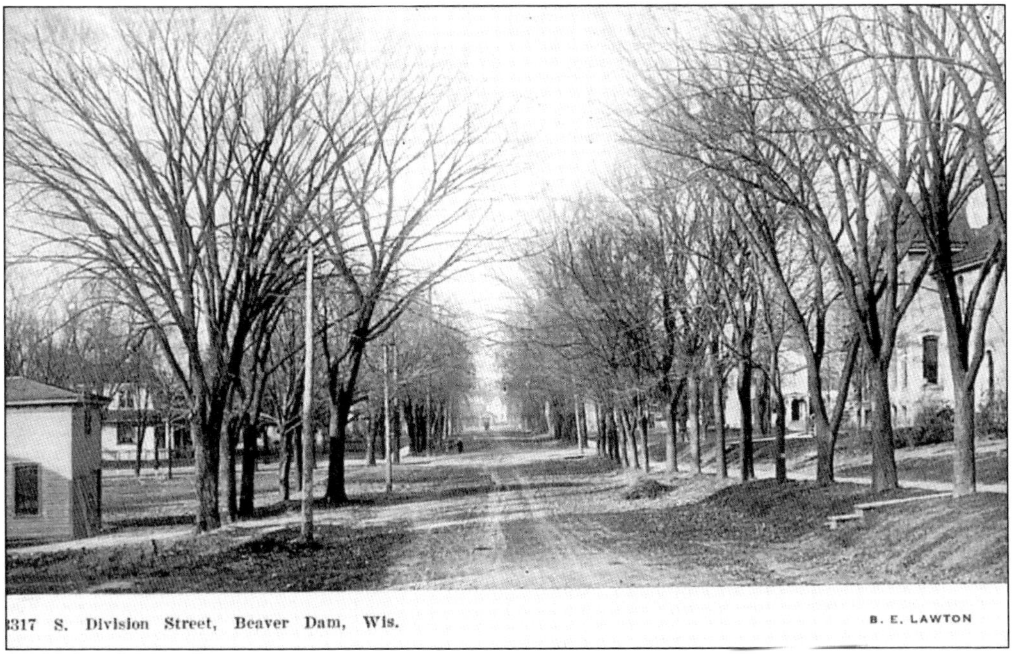

DIVISION STREET. Lincoln Avenue was originally named Division Street. The name was changed around 1928.

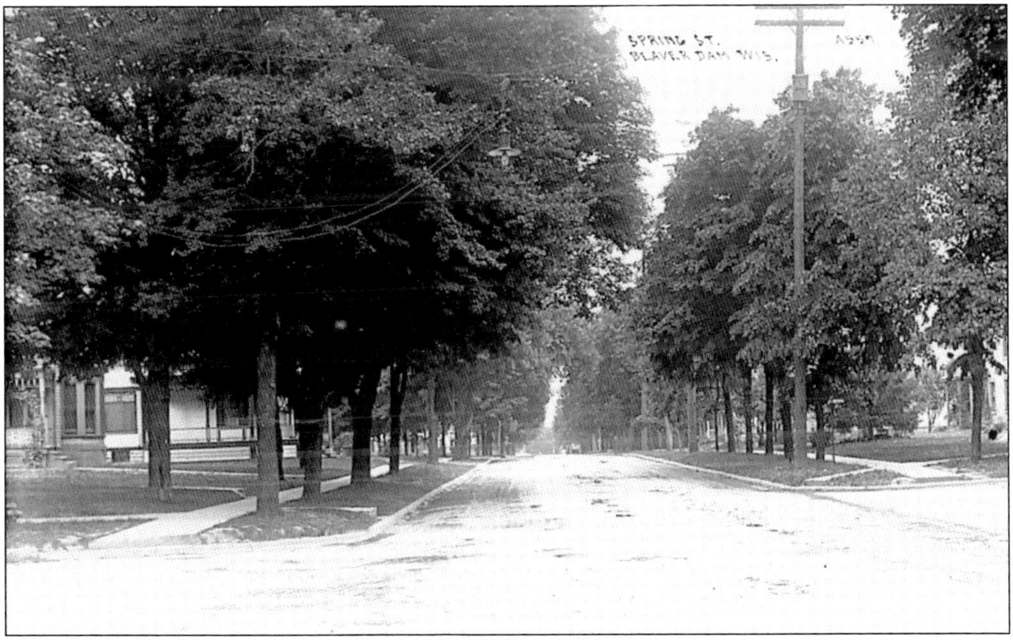

SPRING STREET. The corner of Mackie and Spring Streets is the vantage point for this postcard, which was postmarked in 1918. The second house on the right in the background is the brick home of Civil War hero Horace Patch.

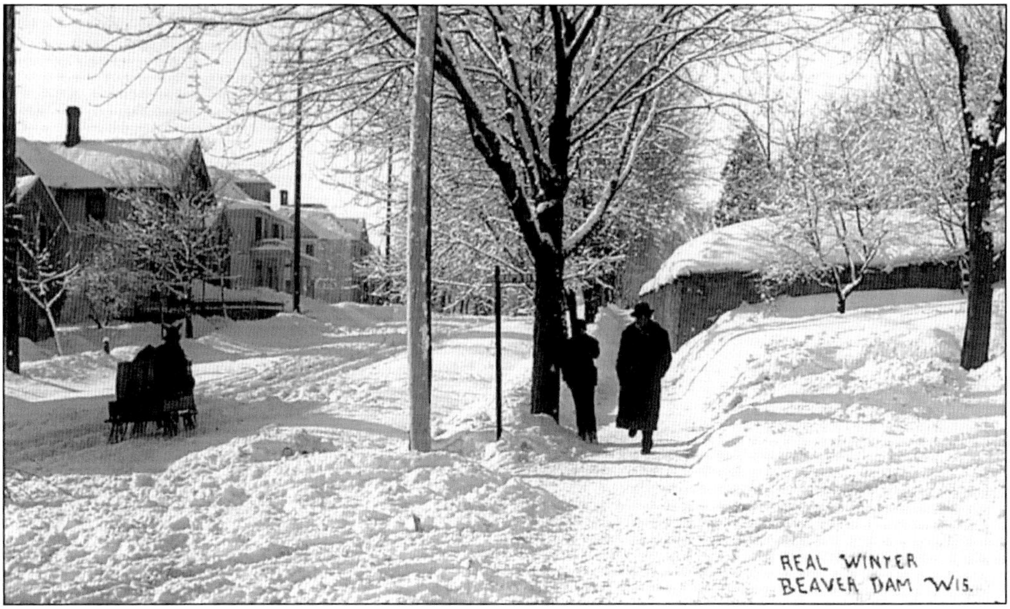

FRONT AND BEAVER STREETS. Entitled "Real Winter," this view looks west up Front Street from the corner by Davison Theatre.

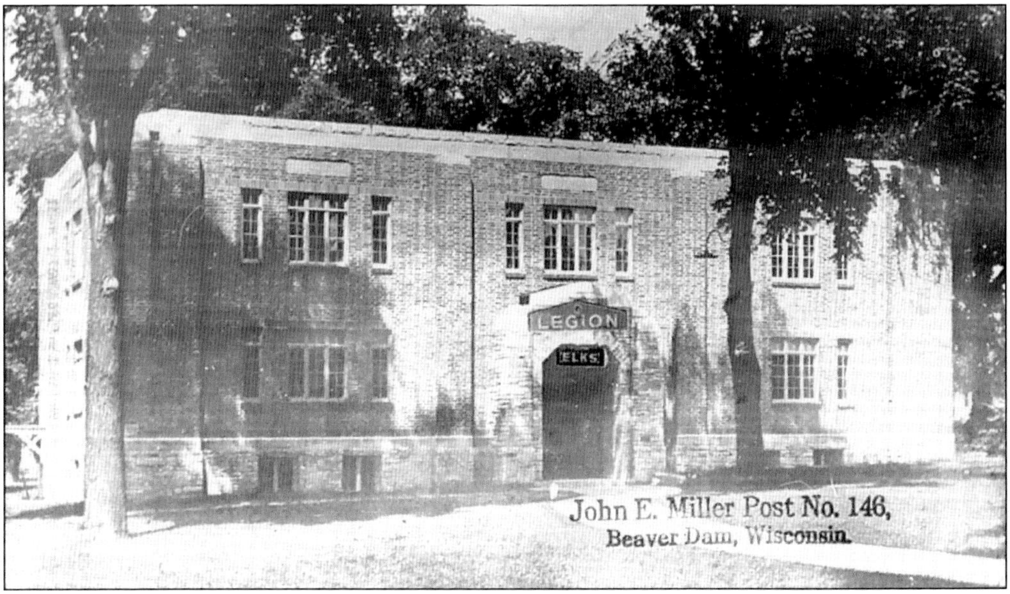

THE LEGION BUILDING. Located at the end of West Front Street in French's Park, the Beaver Dam Legion Memorial Building opened in 1935. The American Legion John E. Miller Post 146 sponsored the legion band, a summer baseball team for high school students, Camp American Legion (a rehabilitation camp), and many other civic projects.

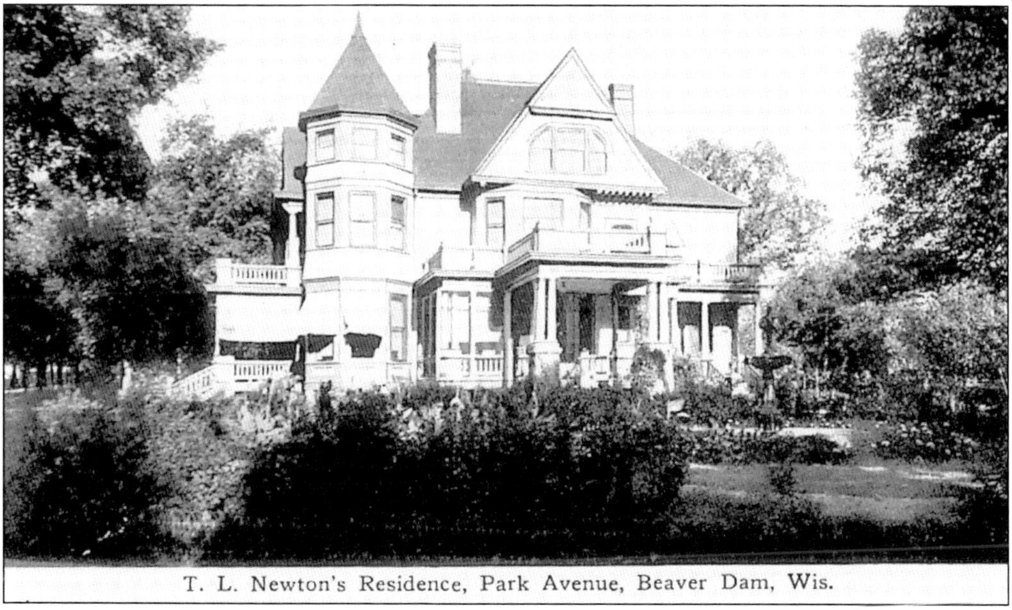

THE T. L. NEWTON MANSION. T. L. Newton was born in Templeton, Massachusetts, in 1827. At the age of 29, he traveled west to Beaver Dam, where he clerked in the A. P. Lawrence grocery. Soon thereafter, he opened his own grocery store. In 1865, he removed himself from the grocery business, opening a general merchandise store. In 1900, the store was sold to Newton's clerks and was incorporated, with Newton retaining stock in the enterprise. Newton's other positions included vice president of the Old National Bank and president of the State Agricultural Society.

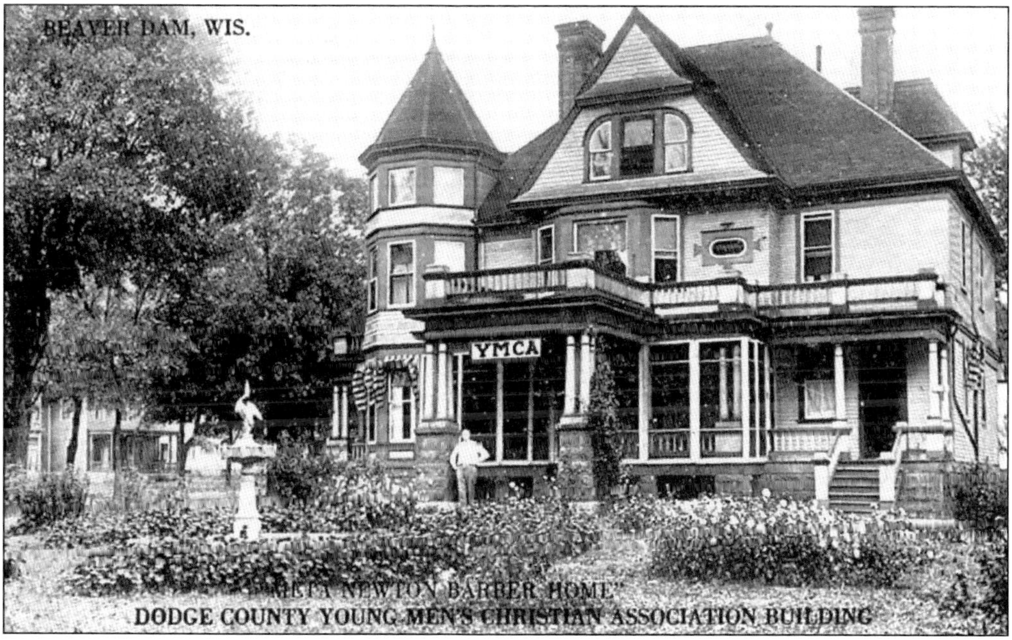

THE META NEWTON BARBER YMCA. T. L. Newton passed away in 1920. He bequeathed his home to the YMCA on the condition that it be named after his daughter, Meta Newton Barber.

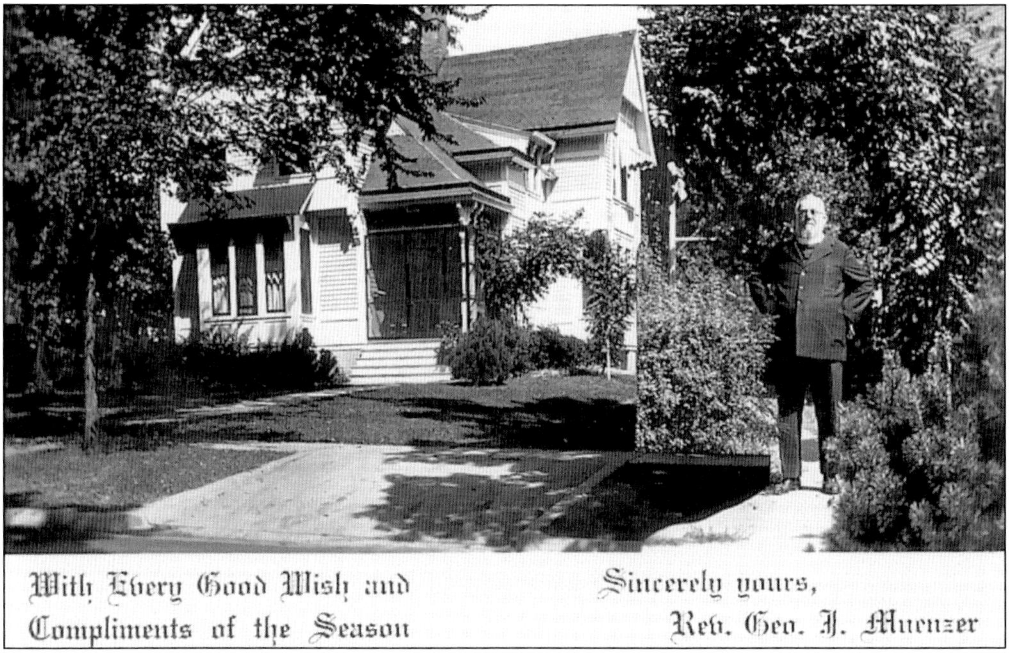

THE REV. GEORGE MUENZER HOME. Rev. George Muenzer served as pastor of St. Peter's Catholic Church from 1909 to 1929. During his term of service, St. Peter's erected a new school and sisters' home. Upon his retirement, Reverend Muenzer remained in Beaver Dam, residing at this home at 212 West Third Street.

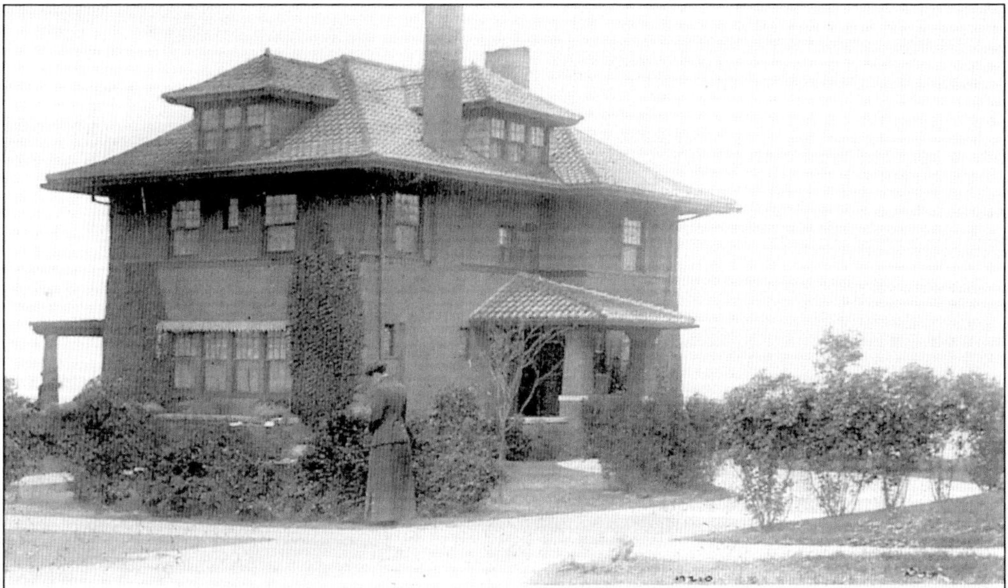

SILAS MCCLURE'S RESIDENCE. Silas McClure arrived in Beaver Dam to serve as secretary of the new manufacturing interest called the Malleable Iron Range Company, more commonly known by the name of its manufactured stove, Monarch. An associate of A. G. Hill and F. W. Rogers, McClure stayed at Monarch until his retirement in 1920. The McClure residence was located at the northwest corner of LaCrosse and Third Streets, just north of the Lamoreaux estate.

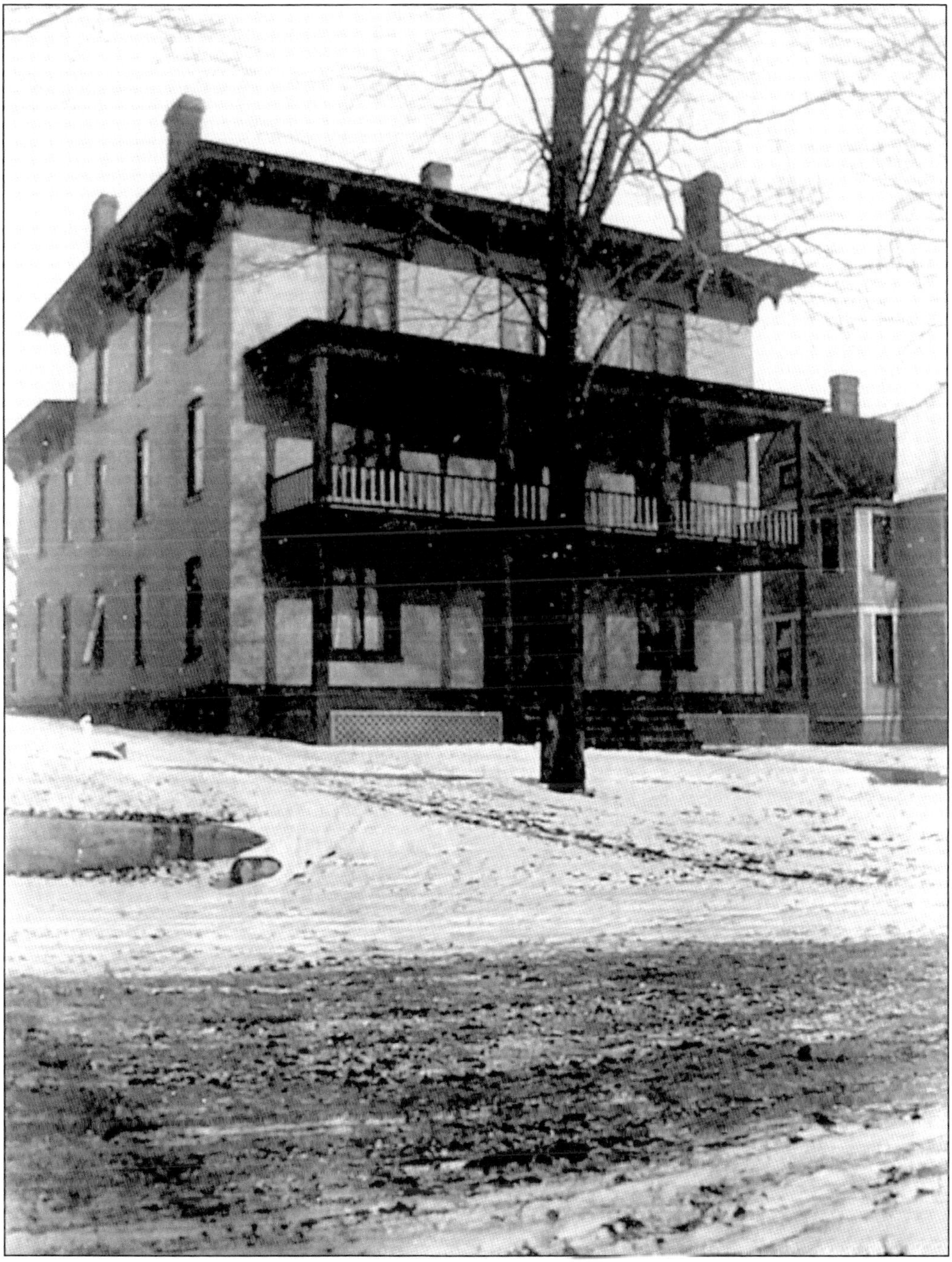

THE BOGERT HOUSE. Located at 212 West Maple Street, the Bogert House was built by R. V. Bogert. Bogert came to Beaver Dam in 1849 to open a general store. He parlayed his business interests into partnerships in the Empire Flour Mills and the Woolen Manufacturing Company. The Bogert estate covered three-quarters of the block and was the first home in the city to have indoor plumbing. A third story was added to the two-story mansion in 1912 to accommodate Beaver Dam's first hospital.

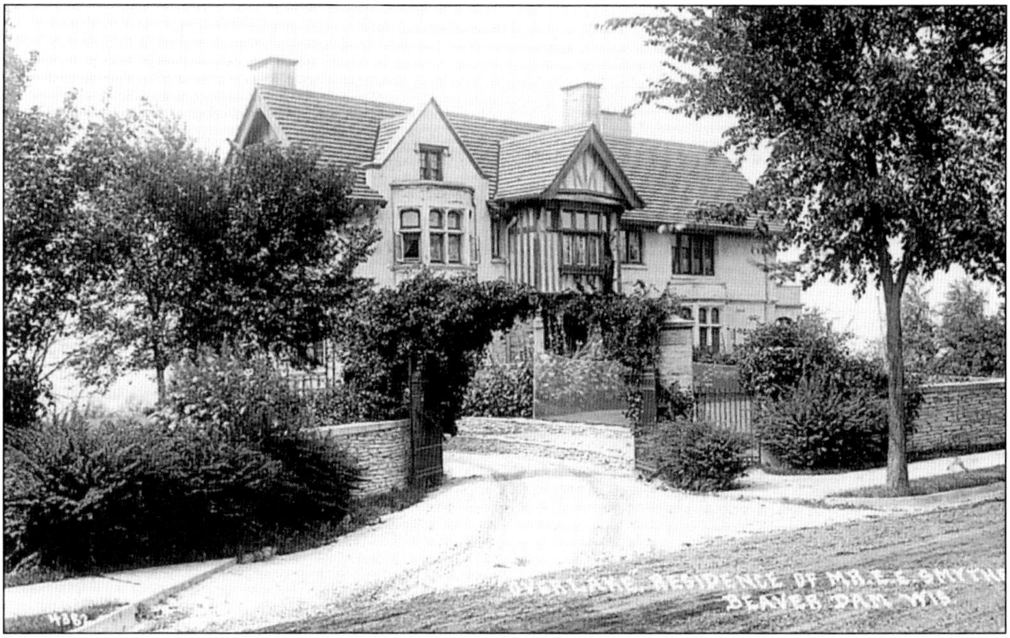

THE LAMOREAUX ESTATE. The Lamoreaux estate was built in 1911 by Percy Lamoreaux. The Lamoreauxs gained their wealth through the Beaver Dam Malleable Iron Company. Business misfortunes forced them to sell the company and the house to Western Malleables. E. E. Smythe, a vice president of Western Malleables, then took up residence at the estate.

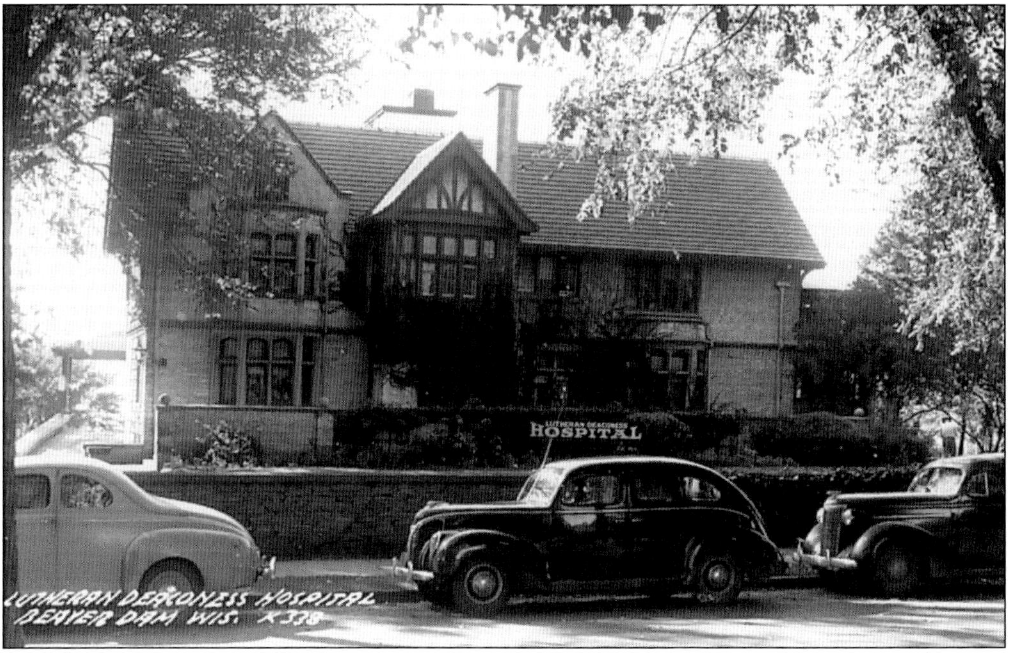

THE LUTHERAN DEACONESS HOSPITAL. In 1922, the Lamoreaux estate was sold by Western Malleables for use as the Lutheran Deaconess Hospital. The building was razed in 1981.

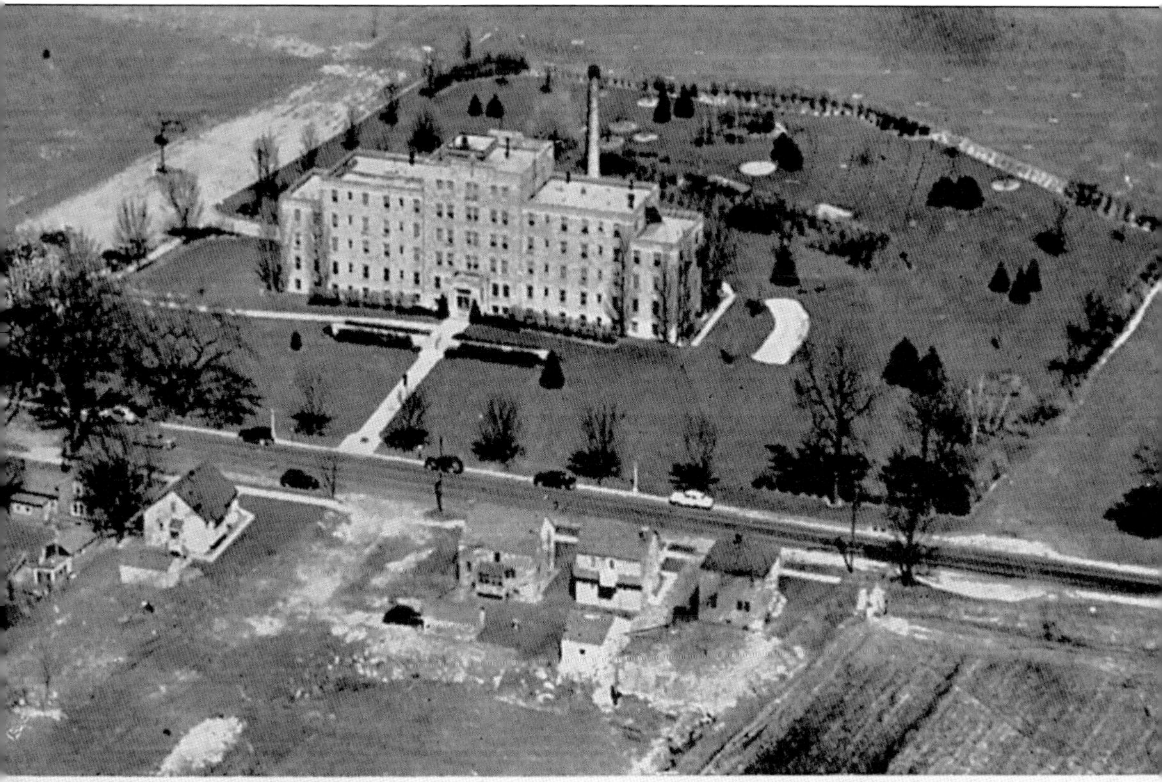

Aerial view of St. Joseph's Hospital, Beaver Dam, Wisconsin.

St. Joseph's Hospital. The 60-bed facility of St. Joseph's Hospital opened in 1938 as a school run by the Sisters of St. Francis. Located on South University Avenue, the hospital received a $2 million addition in 1969. In 1972, the Lutheran Deaconess and St. Joseph's Hospitals merged; Lutheran Deaconess became the Lakeview Long Term Care Unit, and St. Joseph's became Hillside Hospital.

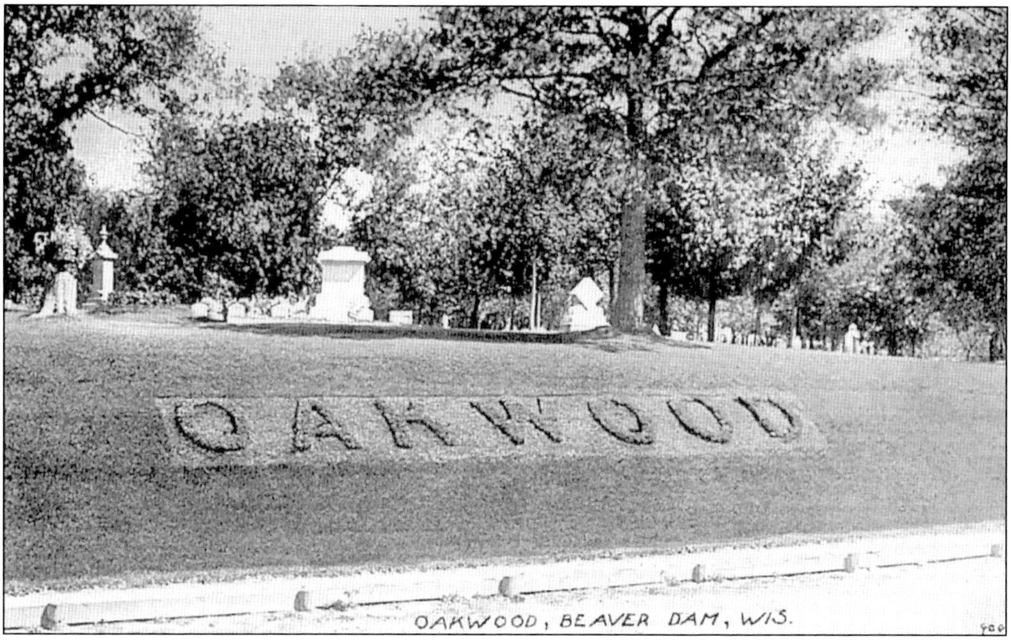

THE OAKWOOD CEMETERY. Ingraham Gould, known for establishing the Gould Nursery, founded the Oakwood Cemetery on June 24, 1869. It is the final resting place for many of the pioneer families of the city.

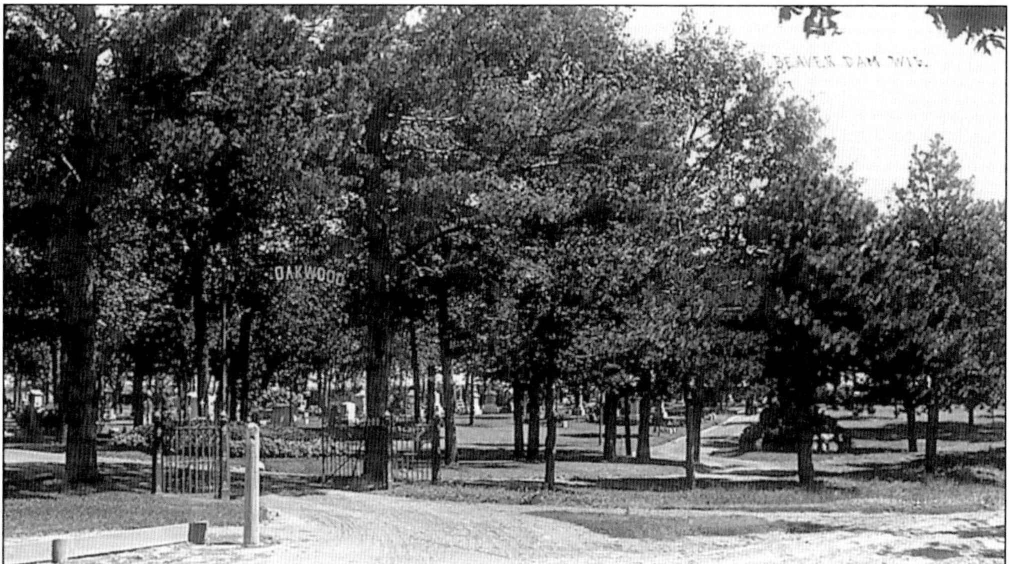

THE OAKWOOD CEMETERY ENTRANCE. The cemetery entrance is located on Park Avenue.

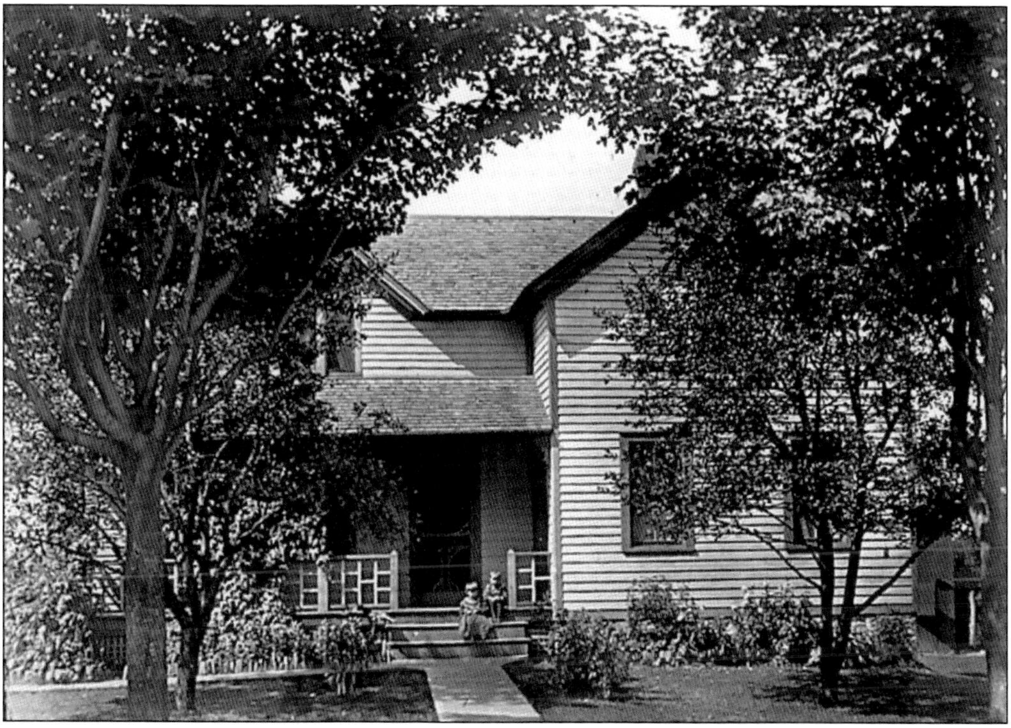

No. 510 Beaver Street. In 1906, the Kodak Company manufactured a pocket camera capable of creating negatives the size of penny postcards. This made it possible for the average Joe to create his own postcard. One of the most prevalent subjects of homemade postcards was the family home. At the time of this photograph, this home was the residence of Frank and John Yagodinski. Later residents included Frederick Schultz, William Schwoch, and L. D. Carlson.

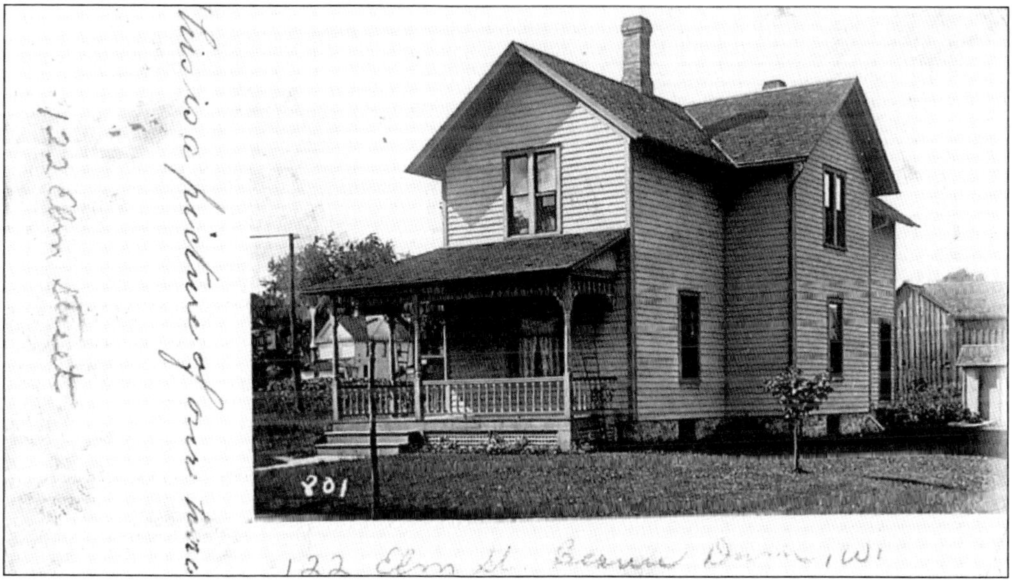

No. 122 Elm Street. At the time this photograph was taken, this home was the residence of George and Martha Gergen. The Gergen family retained possession of this residence well into the late 1950s.

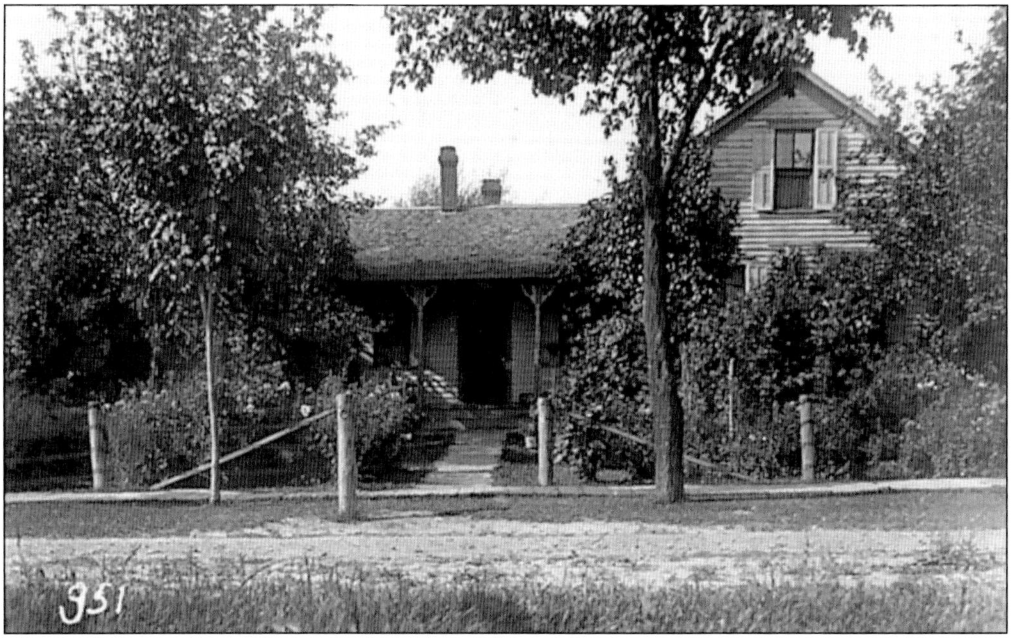

No. 418 West Third Street. When this photograph was taken around 1920, the property was the residence of John Gorzynske. Later, the Albert Biel family owned the home until 1950, and Henry Sheskey was the owner through the 1960s.

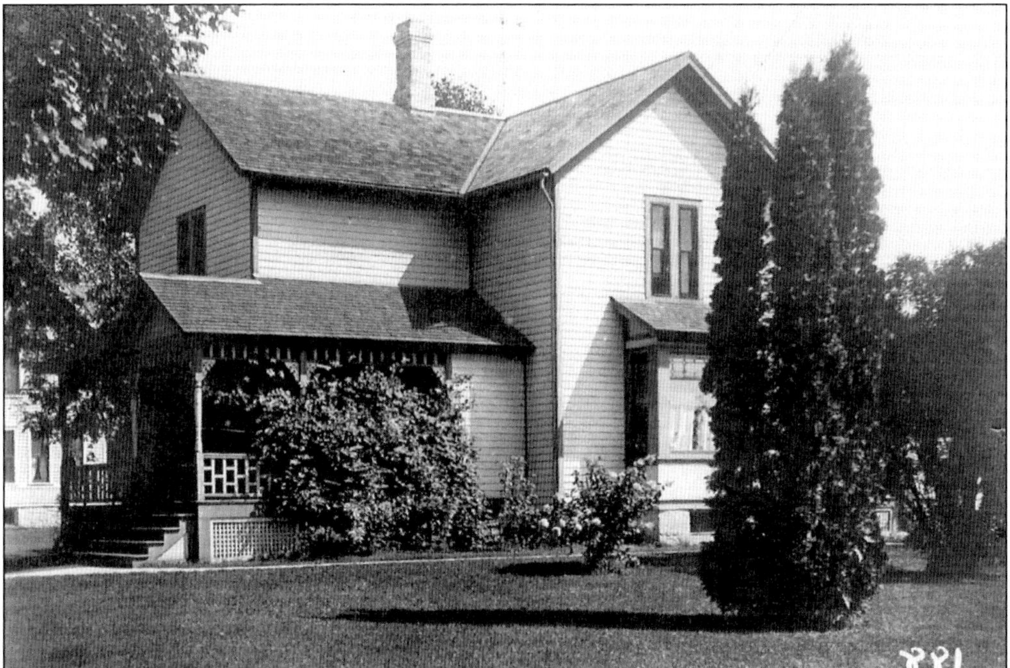

No. 511 North Center Street. This beautiful, modest home possessed such a coveted location that it was razed in the second decade of the 20th century to make room for the F. W. Rogers mansion.

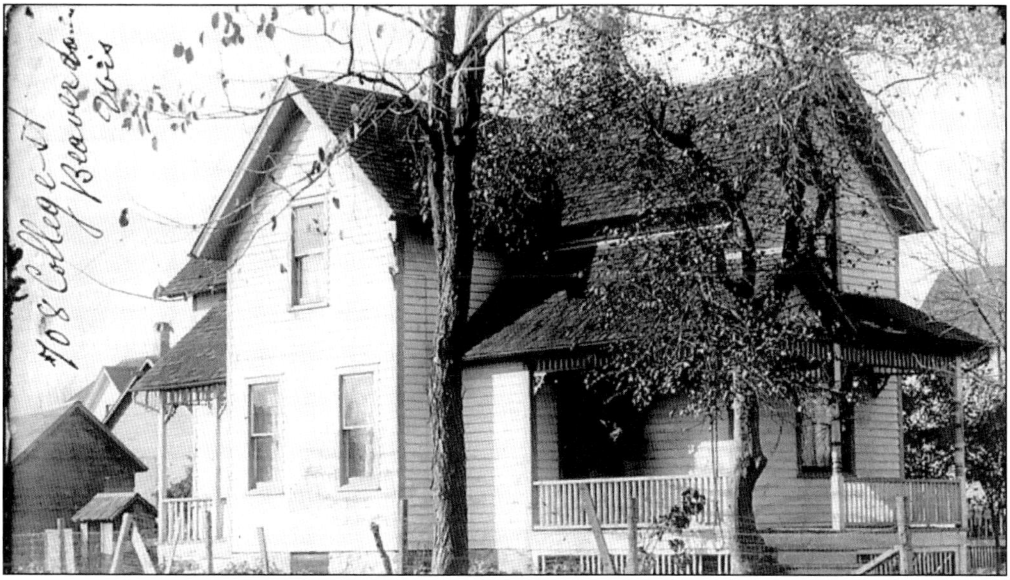

No. 708 College Street. College Street was renamed North University Avenue in 1928. At the time of this photograph, Julius Landry resided here. Later residents included Mary Welch, Albert F. Fowler, and Rodell J. Jacob.

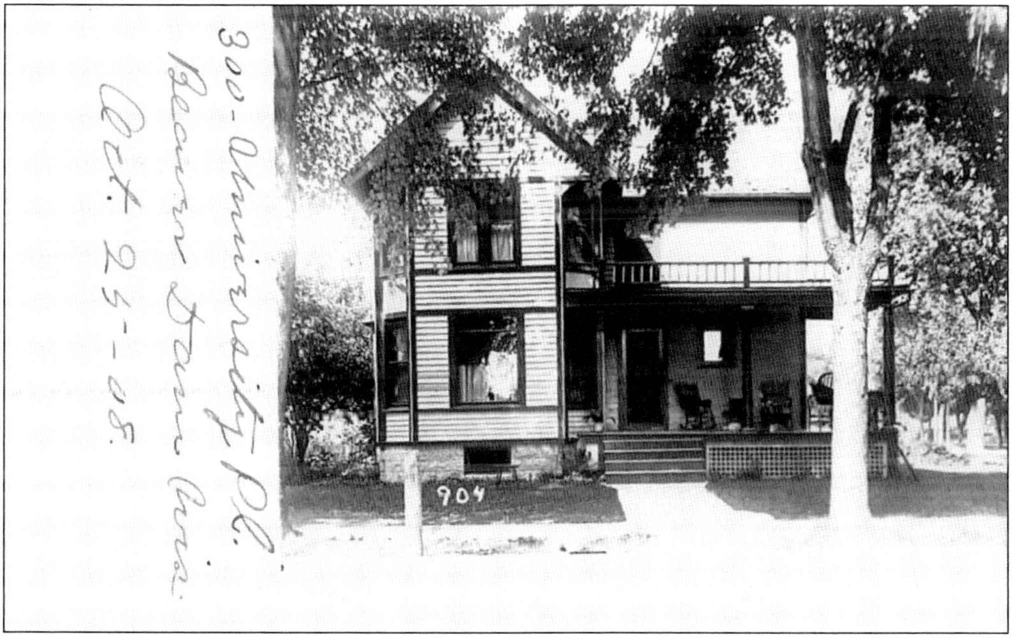

No. 300 University Place. In 1873, the portion of College Street lying south of Park Avenue was renamed University Place. At the time of this photograph, the home was owned by E. E. Parker, assistant postmaster and founding publisher of the *Beaver Dam Daily Citizen*. Subsequent residents included Nettie Coleman, Charles Hron, Walter Roeder, and C. Schimming.

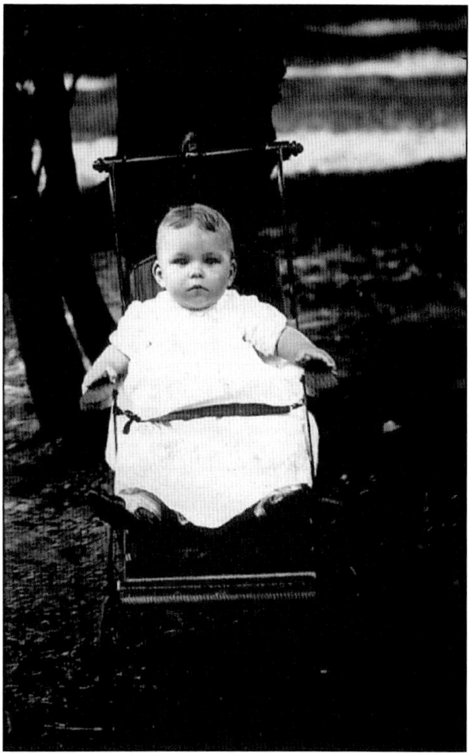

IONA MALLON. Children were another popular subject of private postcards. This photograph was taken at 311 Railroad Avenue at the residence of Charles F. Stehfast, a Malleable Iron Range Company employee. It is presumed that Iona Mallon was a relative or friend. Railroad Avenue was renamed North Lincoln Avenue in 1928.

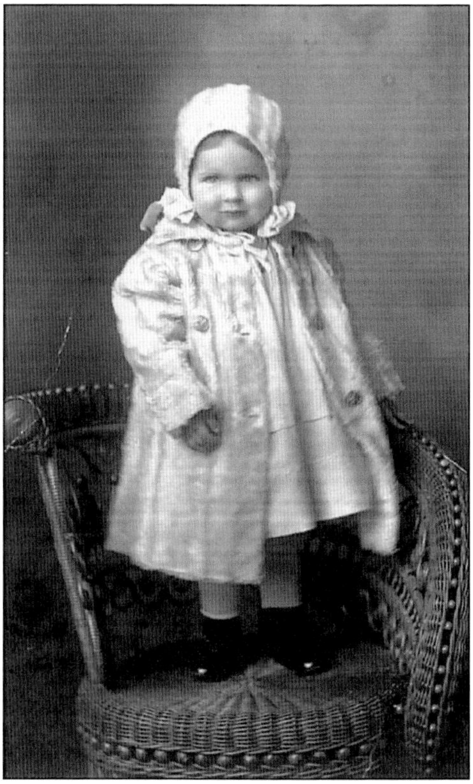

RUTH FEHRER. Taken around 1915, this photograph of Ruth Fehrer was probably snapped at the Fehrer brothers' home at 314 West Mackie Street.

Five

Sixteen Tons and What Do You Get?

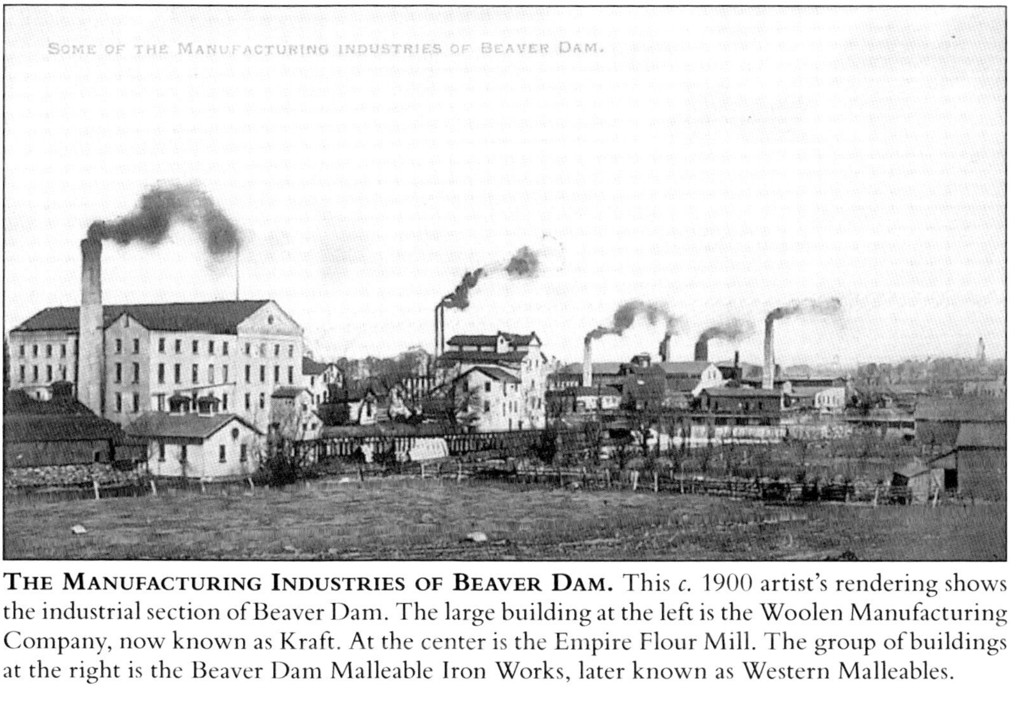

The Manufacturing Industries of Beaver Dam. This *c.* 1900 artist's rendering shows the industrial section of Beaver Dam. The large building at the left is the Woolen Manufacturing Company, now known as Kraft. At the center is the Empire Flour Mill. The group of buildings at the right is the Beaver Dam Malleable Iron Works, later known as Western Malleables.

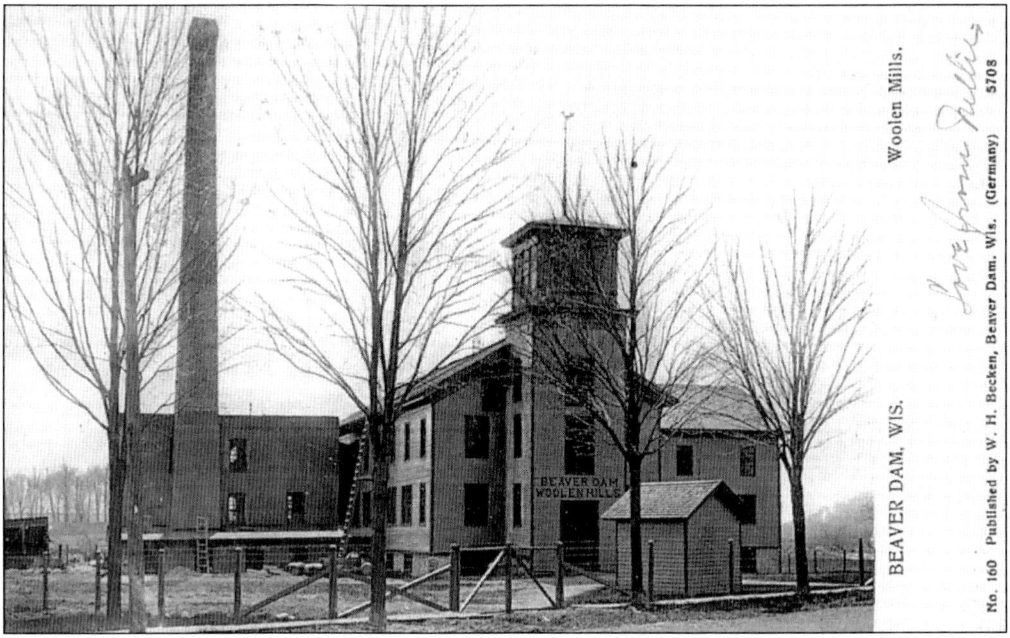

The Beaver Dam Woolen Mills. Situated on the site of Dr. E. H. Hoyt's linseed oil mill, the Beaver Dam Woolen Mills opened in 1866. This structure, also known as the lower woolen mill, was built by the McFetridge brothers and operated under various owners until 1908. From 1913 to 1915, the mill was operated as the Paramount Form Drying Company. The building lay dormant until 1920, when it was razed. The lumber from the building was used to construct local residences.

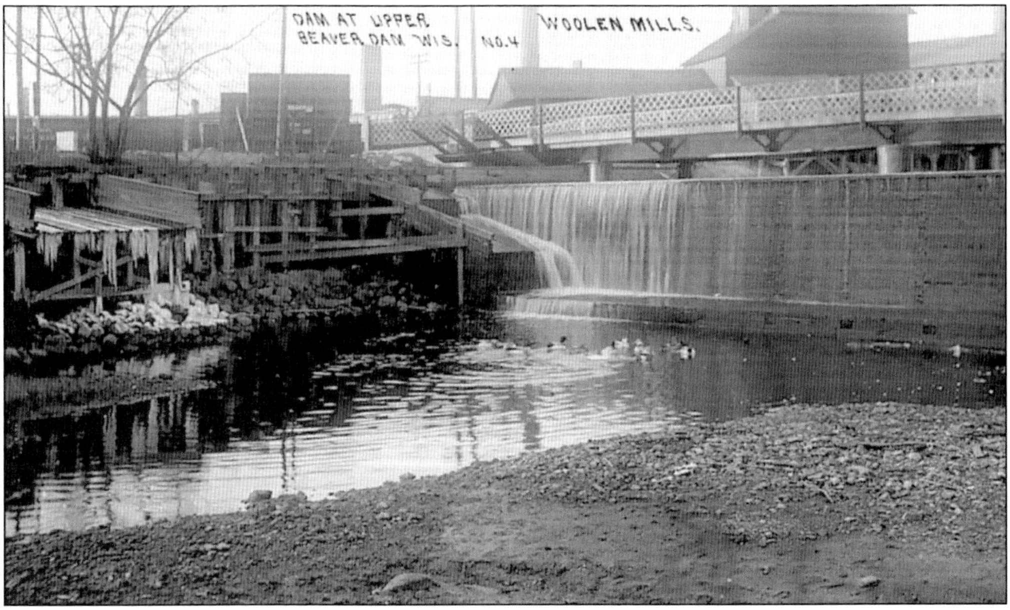

The Dam at the Upper Woolen Mill. This dam, located near the present corner of Mill and Center Streets, served the Woolen Manufacturing Company, also known as the upper woolen mill. Originally, Mill Street did not cross Center Street, and West Mill Street was known as High Street.

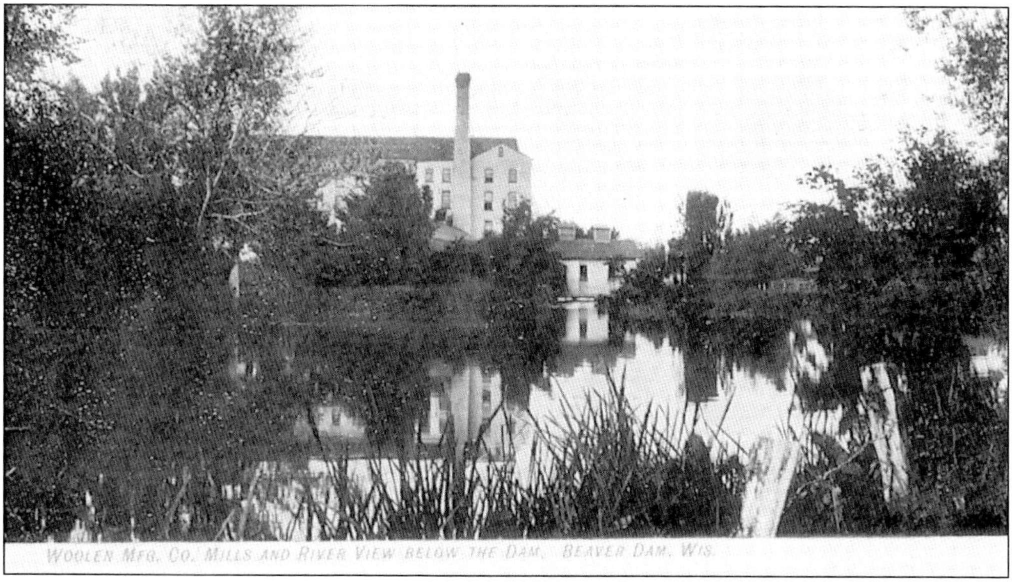

THE WOOLEN MANUFACTURING COMPANY. Founded by town pioneers Abraham Ackerman, S. P. K. Lewis, and G. H. Stewart, the Woolen Manufacturing Company was launched in July 1853. The company was also known as Dodge County Woolen Mills and as Farmers Woolen Mill. In 1867, a new four-story brick mill replaced the original wood structure. In 1882, some 100 citizens worked at the plant, which had been leased to the Beaver Dam Worsted Company. For five years commencing in 1915, the mill operated as a shoe factory for Weyenberg Shoes. It was another five years before the mill reopened as Phenix Cheese.

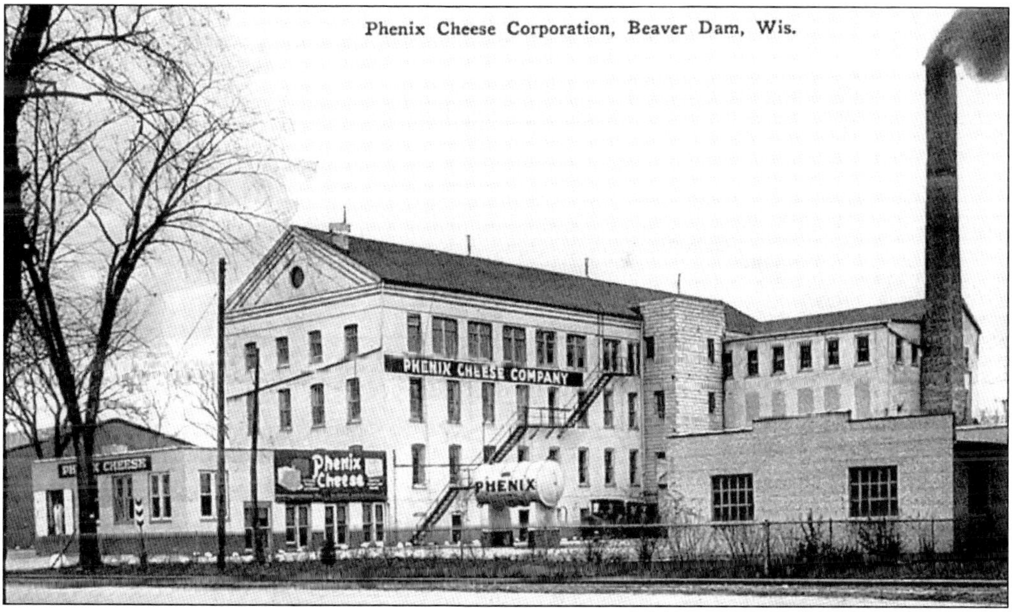

THE PHENIX CHEESE COMPANY. In 1924, the Phenix Cheese Company commenced activity at the upper mill. Initially serving 40 dairies, the enterprise had expanded service to 250 area farmers by 1991. The plant specializes in processing soft cream cheeses. During its 81 years of operation, the plant changed its name to Kraft–Phenix Cheese Company and ultimately to just Kraft.

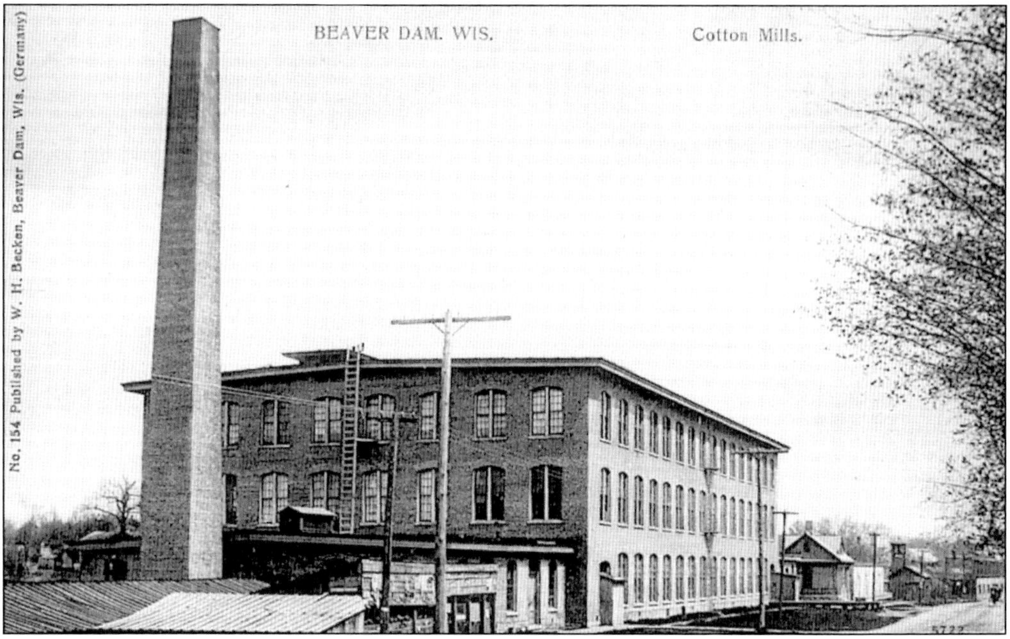

THE COTTON MILL. The Beaver Dam Flouring Mills, owned by David Drake and later by Moses Ordway, transacted business for the first time in 1847. After fires in 1870 and 1881, the flouring business was abandoned in favor of a cotton mill. The cotton mill, specializing in cotton sheeting, began operation the first week of April 1882. The mill closed 24 years later due to southern competition.

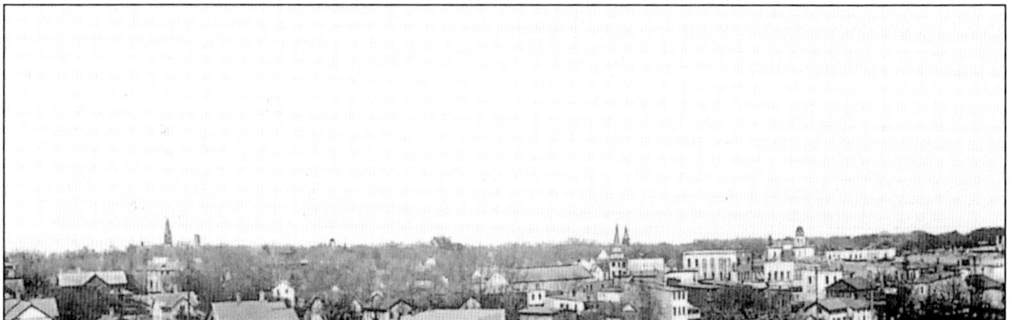

THE VIEW FROM THE COTTON MILL ROOFTOP, LOOKING DOWNTOWN. This 1900s view, looking toward downtown from atop the cotton mill, shows various landmarks of the era. To the right, just east of the cotton mill on Madison Street, is the first city firehouse. Directly to the east along the river is the Beaver Street Bridge.

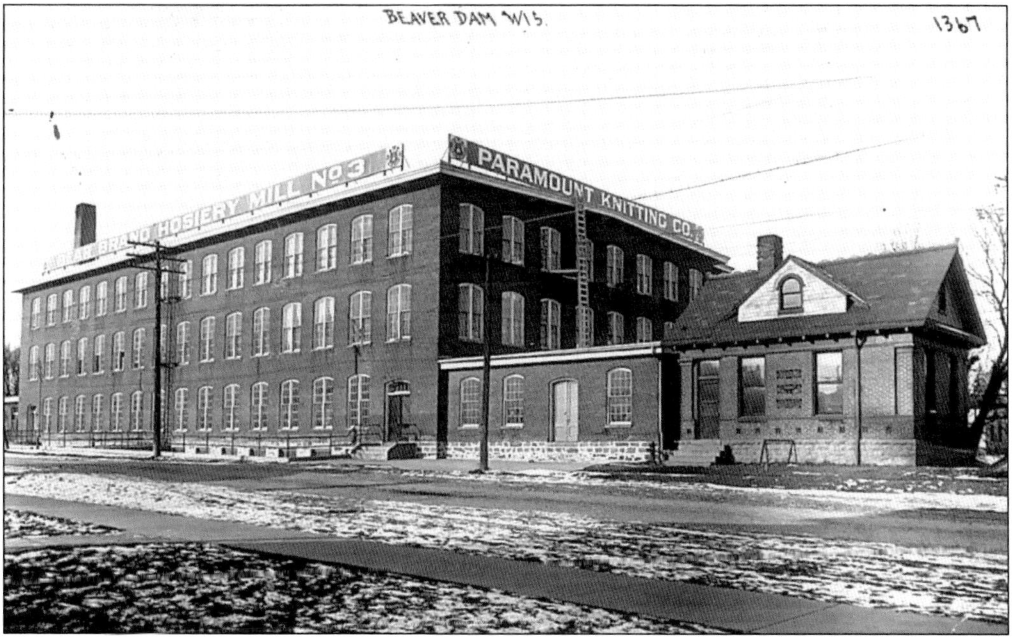

PARAMOUNT KNITTING AND BEAR BRAND HOSIERY COMPANY. The cotton mill had lain vacant for five years when Henry Pope purchased it in 1911 and relocated his Paramount Knitting and Bear Brand Hosiery Company to the site. After an extensive $250,000 renovation, the plant prospered for a quarter of a century, employing over 500 laborers. But a misguided labor force's reliance on union assurances forced a strike in 1934. Pope, having provided employment to his workforce during the very depths of the Depression, was given little choice other than to relocate to Indiana.

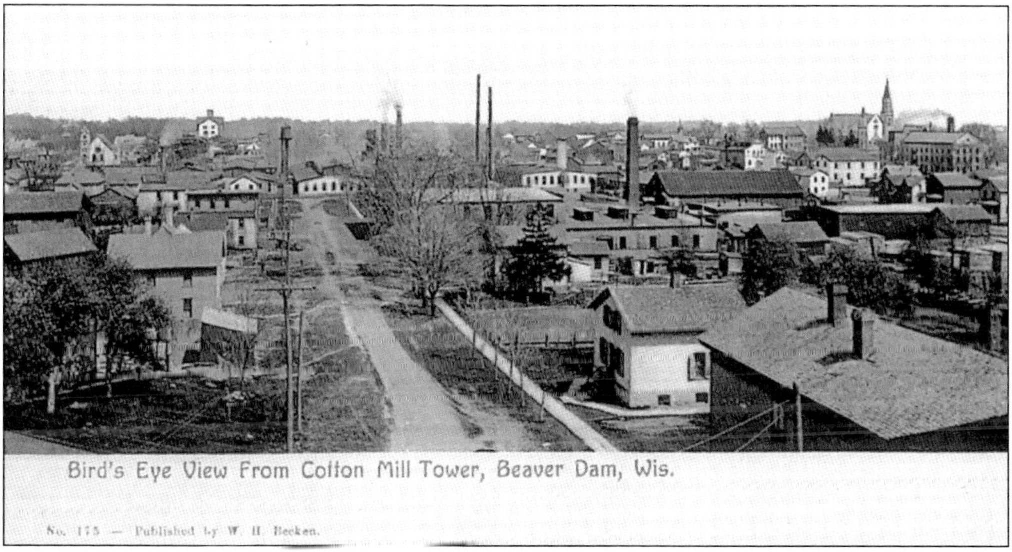

THE VIEW FROM THE COTTON MILL ROOFTOP, LOOKING DOWN ROWELL STREET. This 1900s view from atop the cotton mill shows various landmarks of the era. To the far right is the upper woolen mill, and behind the mill is St. Peter's Church. Down Rowell Avenue, the J. S. Rowell Manufacturing Company is visible, and beyond that is the Beaver Dam Malleable Iron Works, also known as Western Malleables.

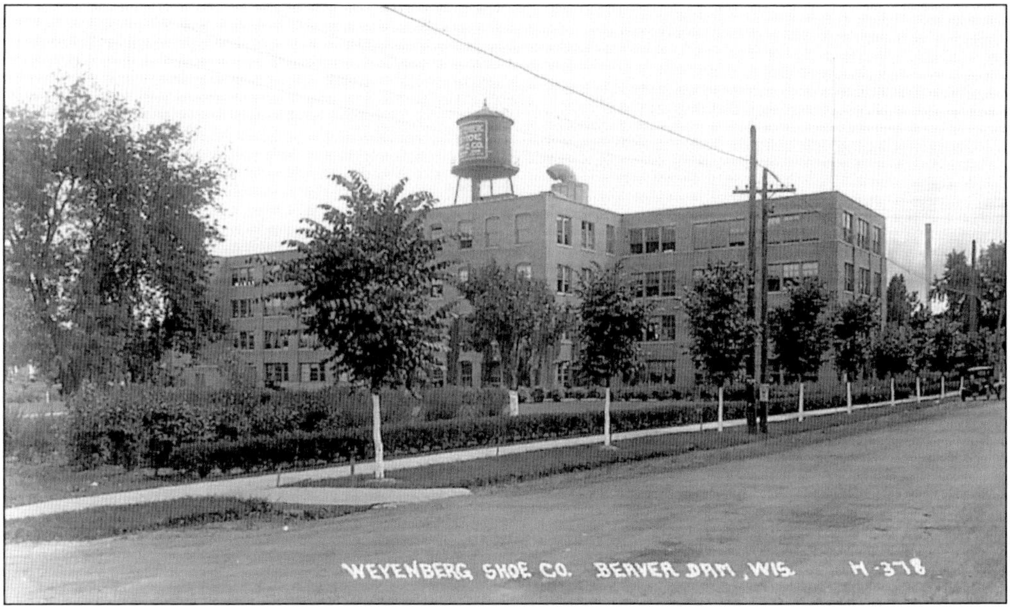

THE WEYENBERG SHOE MANUFACTURING COMPANY. In 1914, the Weyenberg Shoe Manufacturing Company retooled the upper woolen mill to manufacture shoes. The enterprise soon proved so successful that a brand-new plant was required. With the city's aid, Weyenberg's new facility opened in the spring of 1919 at the North Spring Street site shown above. The shoe factory was placed on National Register of Historic Places in 2000.

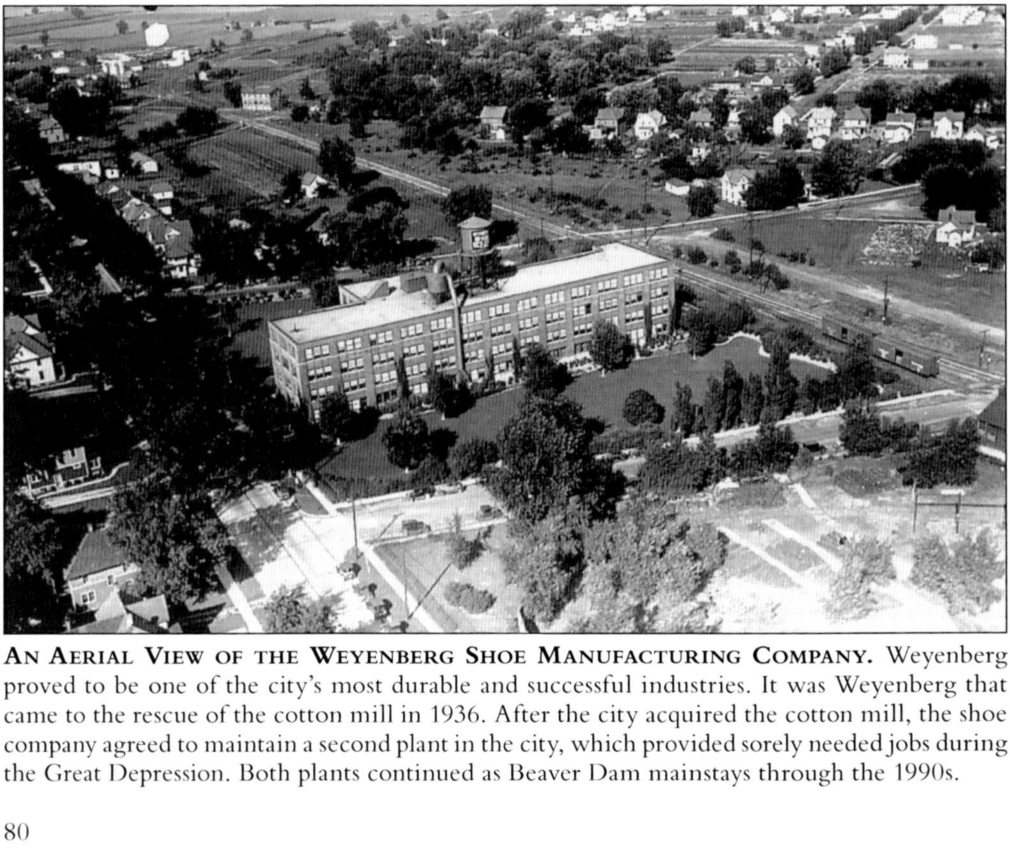

AN AERIAL VIEW OF THE WEYENBERG SHOE MANUFACTURING COMPANY. Weyenberg proved to be one of the city's most durable and successful industries. It was Weyenberg that came to the rescue of the cotton mill in 1936. After the city acquired the cotton mill, the shoe company agreed to maintain a second plant in the city, which provided sorely needed jobs during the Great Depression. Both plants continued as Beaver Dam mainstays through the 1990s.

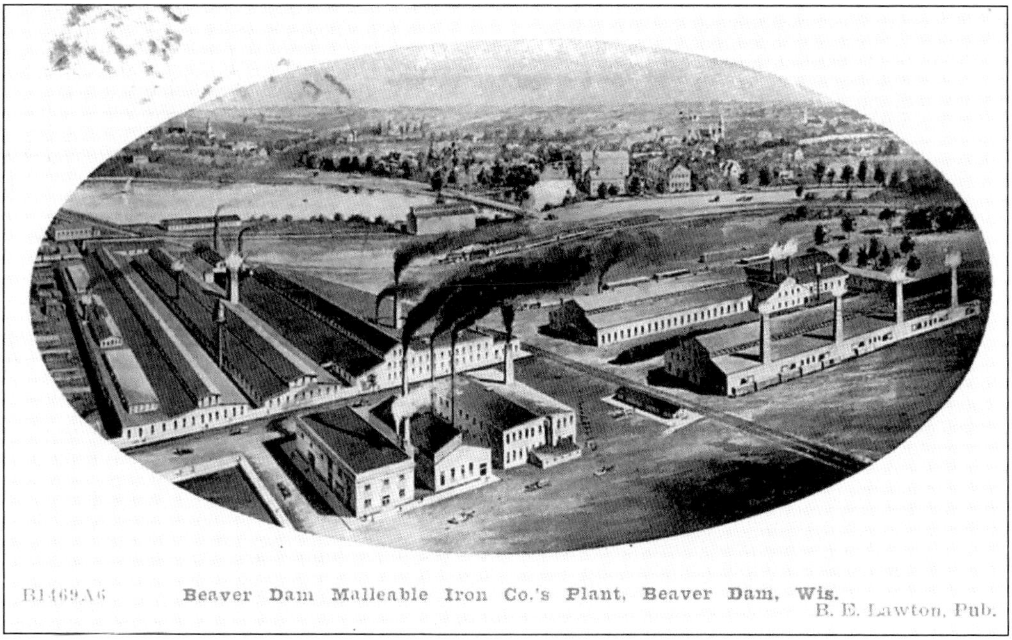

THE BEAVER DAM MALLEABLE IRON WORKS. In 1892, the Beaver Dam Business Men's Association entered into an agreement to create the Beaver Dam Malleable Iron Works. The company manufactured gray iron, malleable iron, and brass casings at its site on Center Street at the Rowell Street connection. The first orders arrived from the J. S. Rowell Manufacturing Company and from Van Brunt of Horicon (now John Deere).

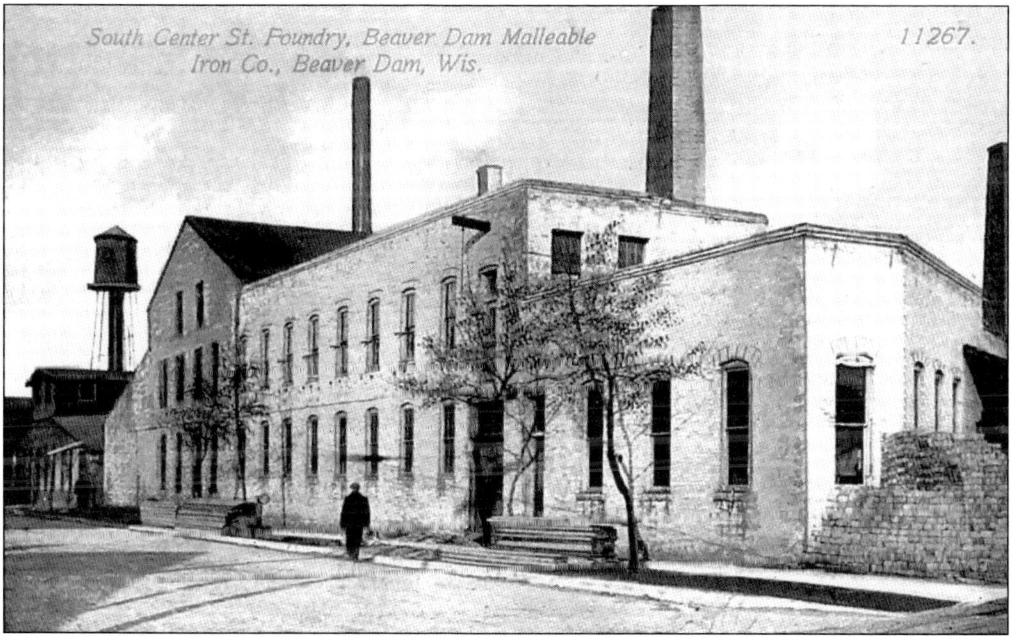

THE SOUTH CENTER STREET FOUNDRY. In 1900, the Beaver Dam Malleable Iron Works factory suffered substantial damage from fire. The company, now called the Beaver Dam Malleable Iron Company, rebuilt and expanded with structures composed of durable brick.

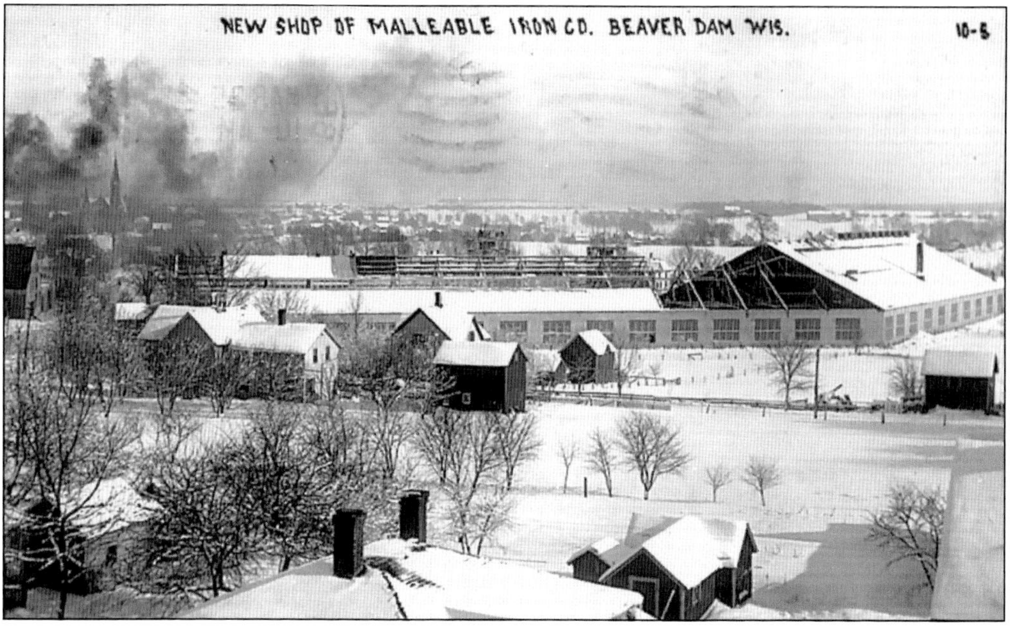

The New Shop at the Beaver Dam Malleable Iron Company. The company prospered and soon outgrew its facility. In 1906, an additional plant was added on Elm Street. In 1909, a new foundry was erected on South Street.

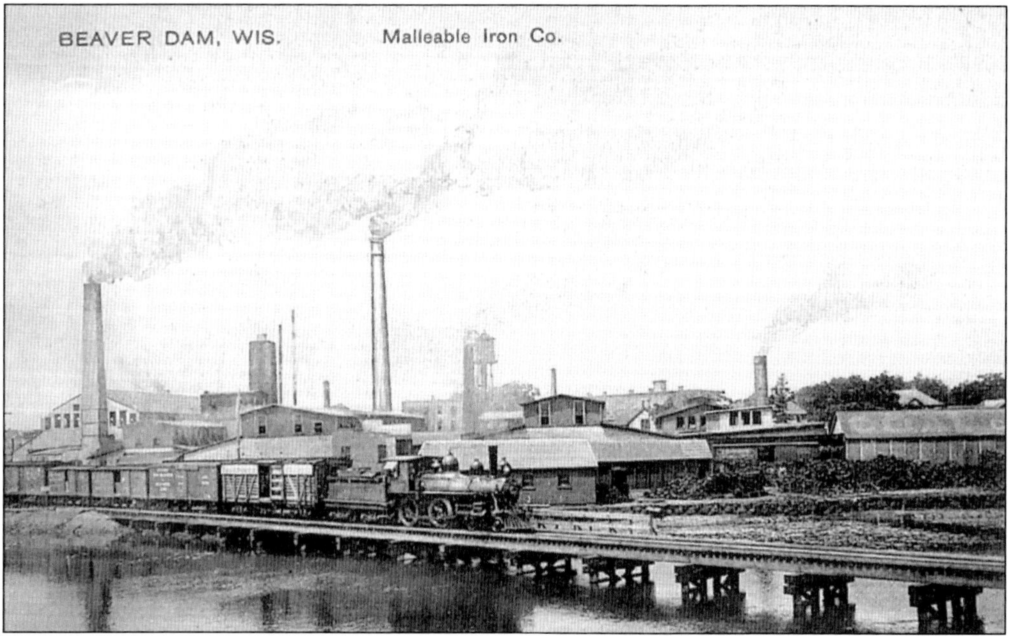

The Railroad Bridge over the Downtown Millpond. Thanks to part-owner Lawrence Fish's connections with the Milwaukee Road, the Beaver Dam Malleable Iron Company became one of the nation's largest malleable iron concerns, employing over 1,500 people. With increased competition in the 1920s, the plant was forced to reorganize under the name Western Malleables. The plant ceased production in 1936.

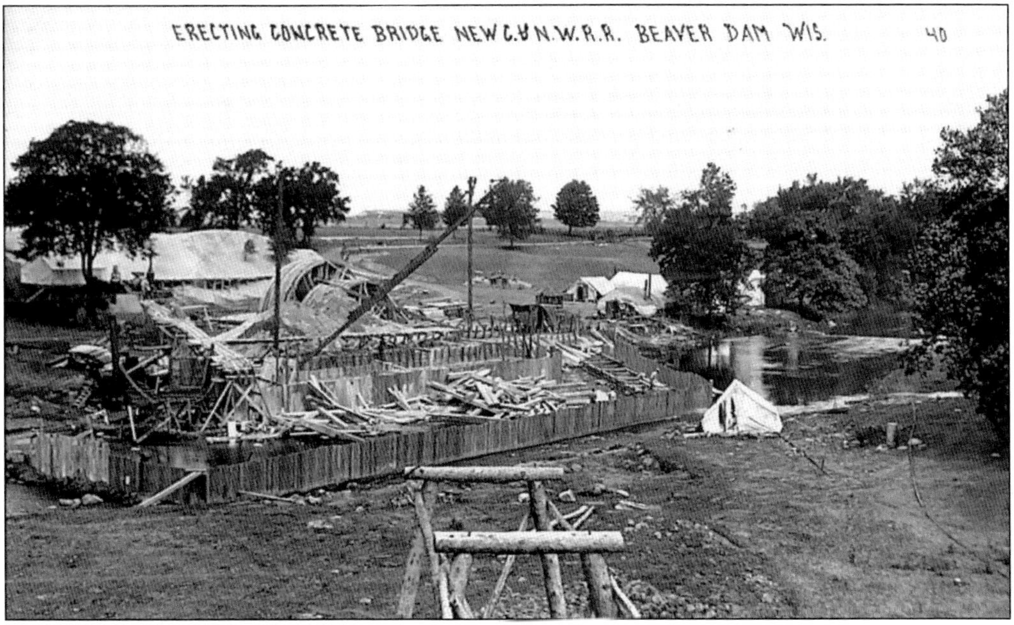

THE CHICAGO AND NORTHWESTERN RAILROAD BRIDGE. The railroad played a significant role in the development of Beaver Dam. The LaCrosse & Milwaukee Railroad steamed into the North Spring Street depot for the first time in 1856. But work on the Chicago and Northwestern Railroad lines, pictured here, occurred later; track was laid for this line in 1910 just outside the city in south Beaver Dam.

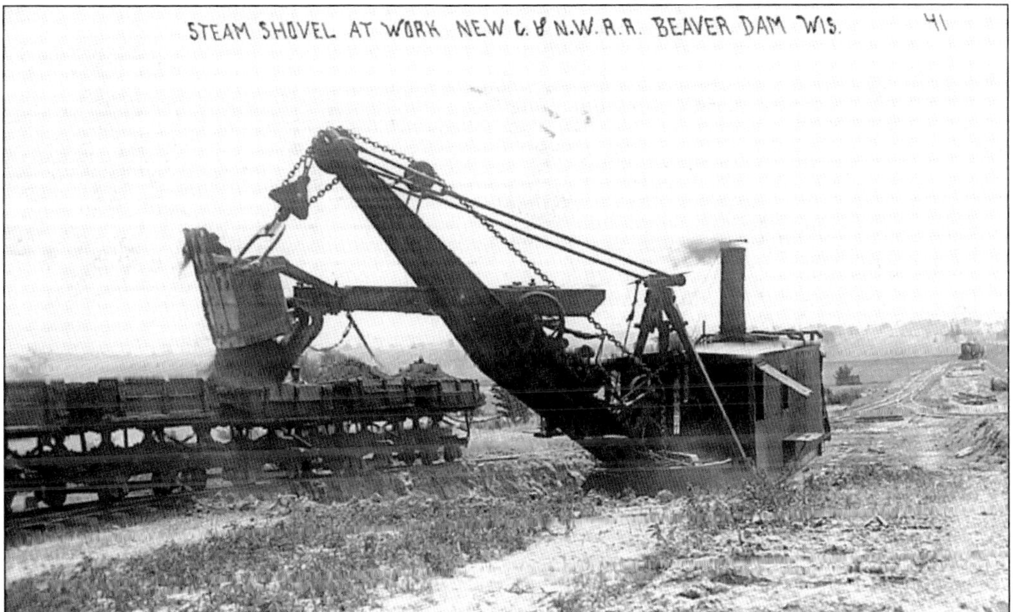

A CHICAGO AND NORTHWESTERN RAILROAD STEAM SHOVEL. The Chicago and Northwestern Railroad provided both freight and passenger service beginning on December 15, 1911. Engine No. 400 averaged speeds of over 70 miles per hour as it traveled across the country.

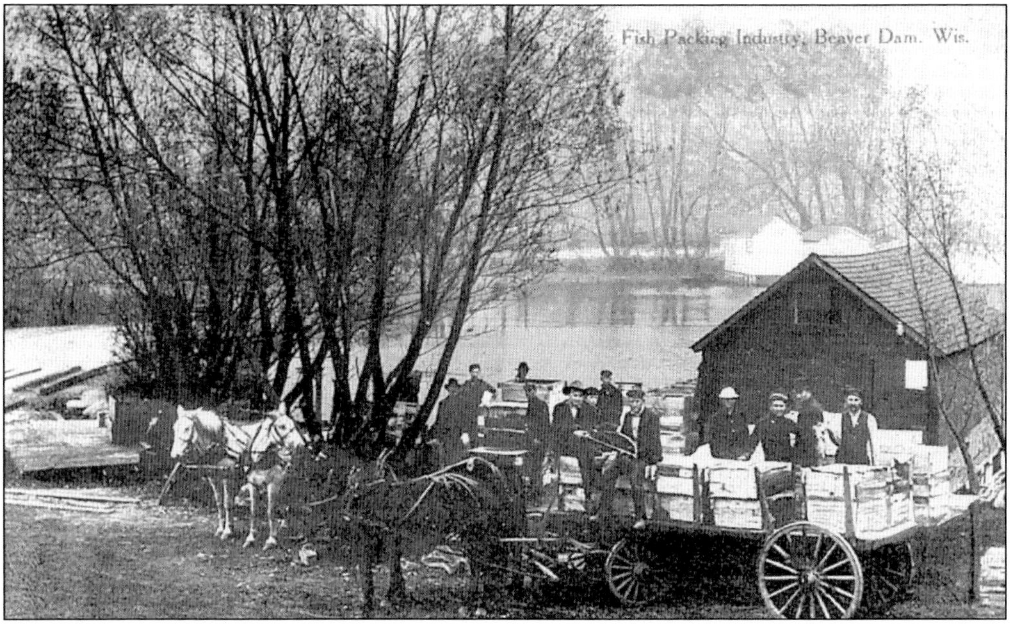

THE FISH PACKING INDUSTRY.

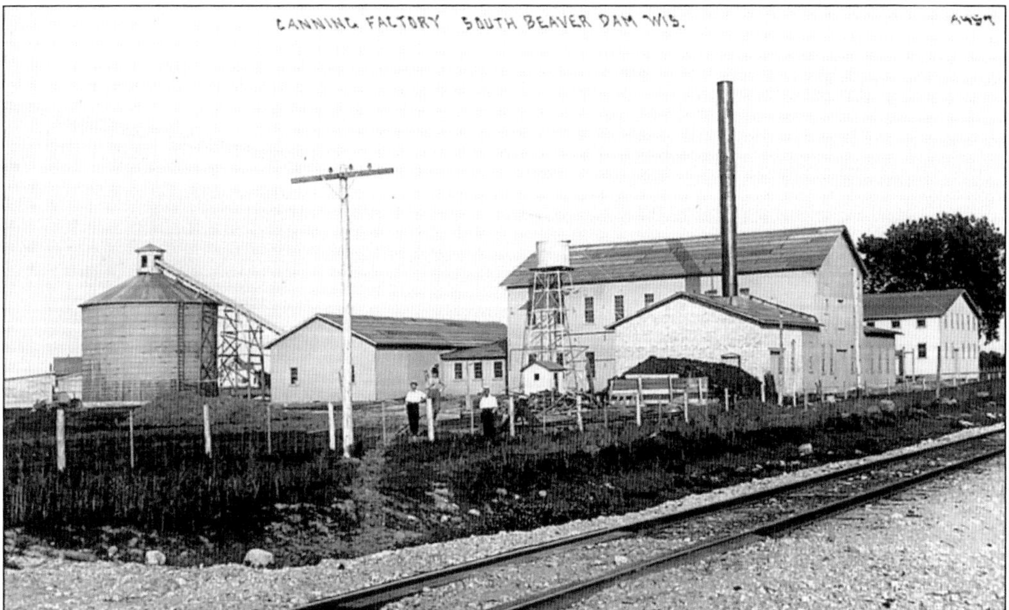

THE CANNING FACTORY. The Beaver Canning Company commenced business in 1912. The south Beaver Dam entity initially canned peas, expanding in later years to packing corn.

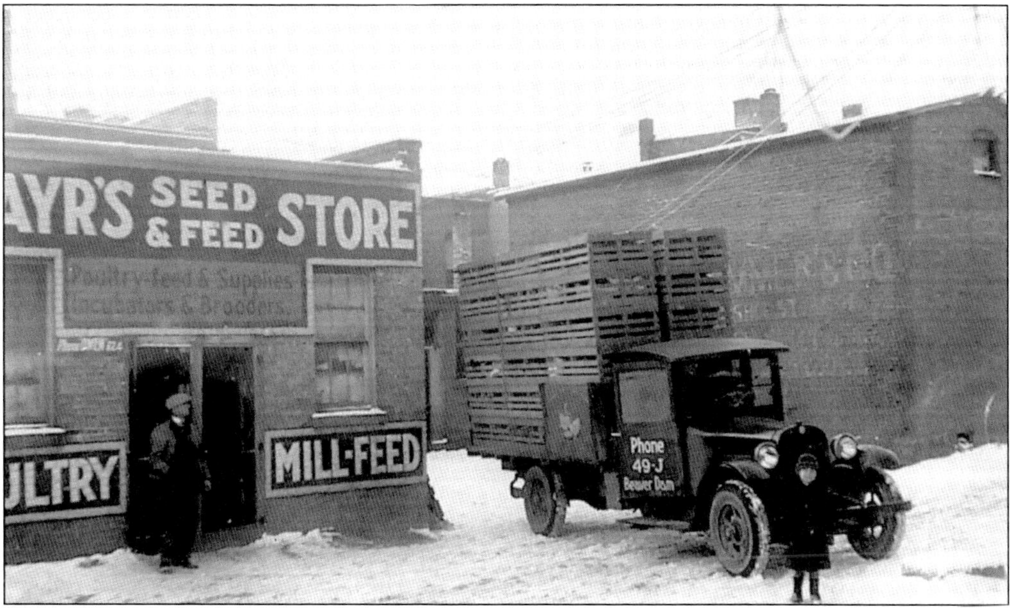

MAYR'S SEED AND FEED. I. K. Mayr arrived in Beaver Dam in 1924 and entered into the grocery and seed-and-feed business. He rapidly acquired various facilities throughout the city, including a storefront at 134 Front Street and elevators at Beaver Dam Junction, Mill Street, and South Spring Street.

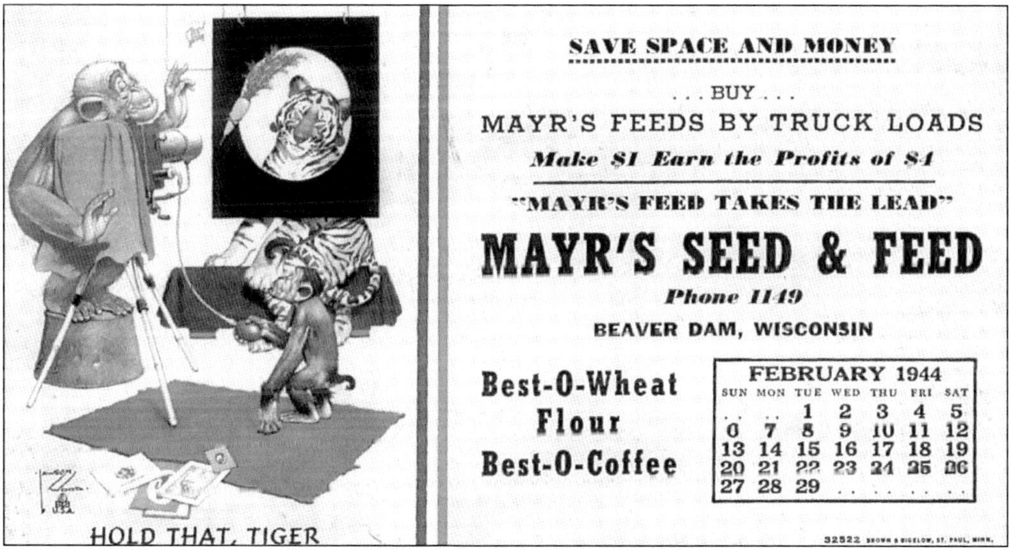

A MAYR'S SEED AND FEED ADVERTISEMENT. This Mayr's Seed and Feed postcard is reminiscent of the initial postcards of the late 19th century, which were used to advertise business enterprises.

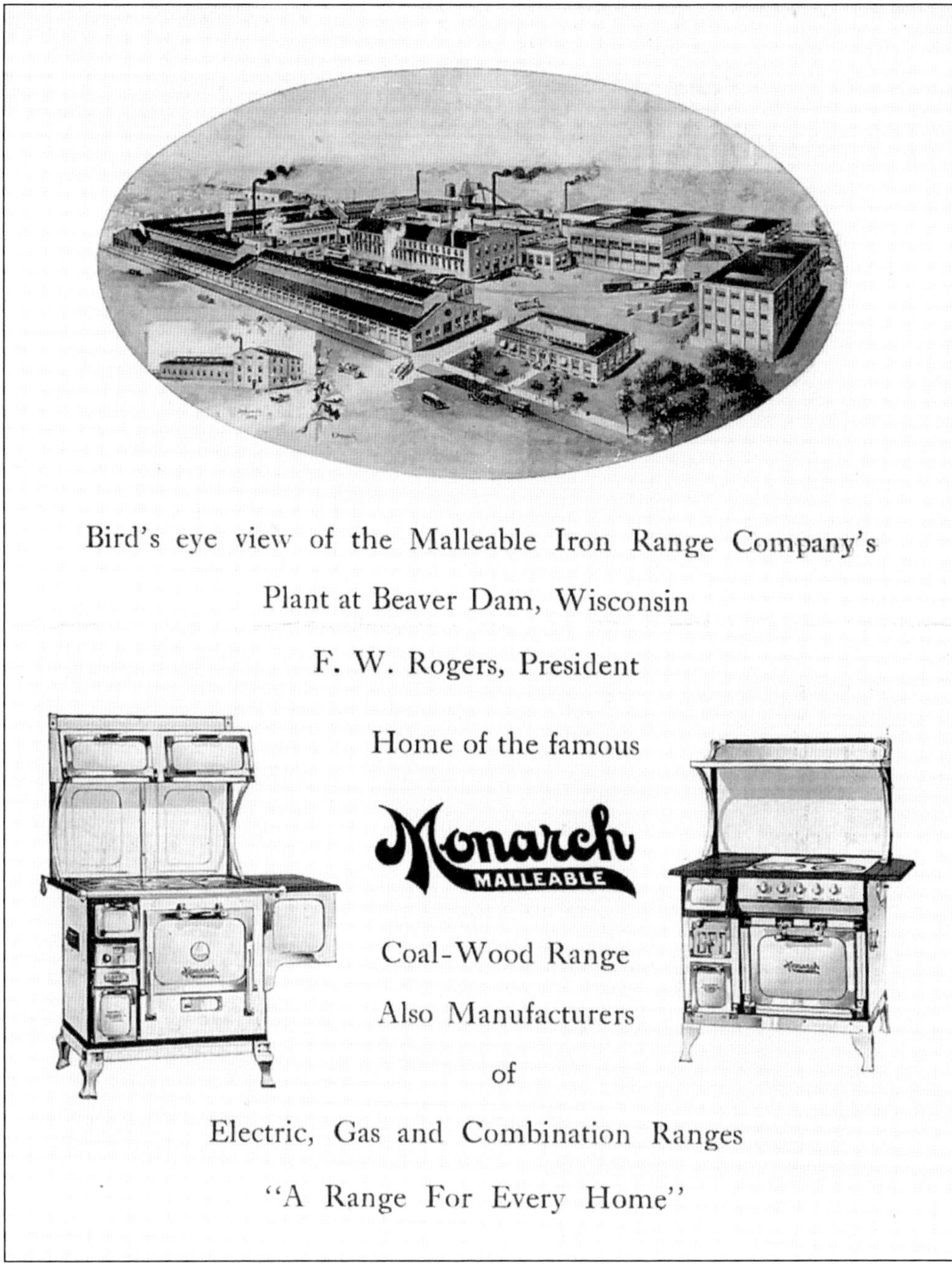

THE MALLEABLE IRON RANGE COMPANY. The triple-sided postcard appearing on these two pages is more of an advertisement for F. W. Rogers than for the Malleable Iron Range Company. In 1900, the Dauntless Manufacturing Company became indebted to the Beaver Dam Malleable Iron Works. Dauntless failed to make good on its debt, and its assets were transferred to the Beaver Dam Malleable Iron Works. Two years later, the corporation was renamed the Malleable Iron Range Company, makers of the reliable Monarch electric and Paramount gas stoves.

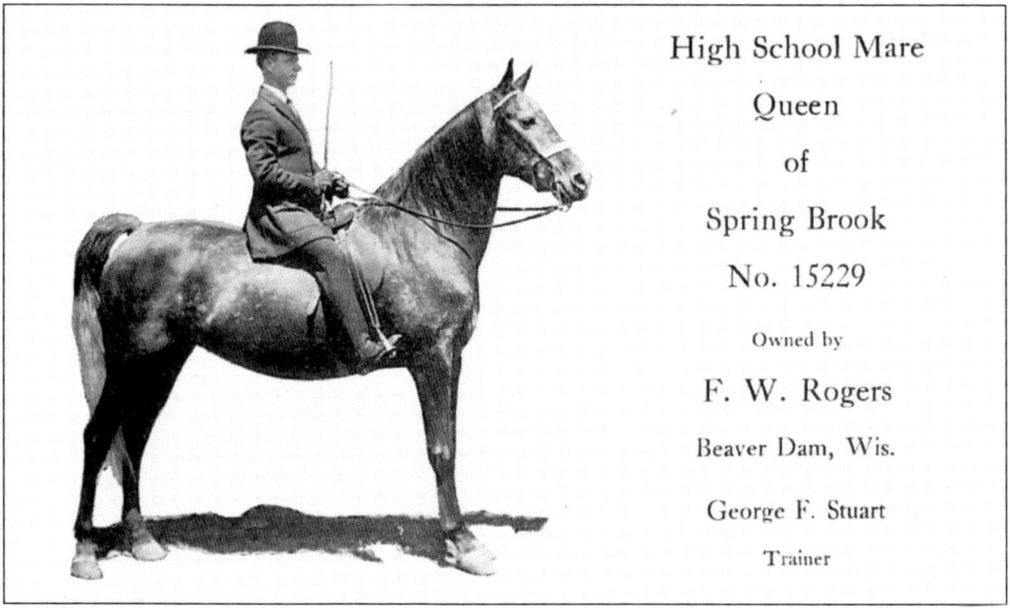

F. W. Rogers. F. W. Rogers joined the Malleable Iron Range Company in 1902 and spent the next 12 years crisscrossing the country as a salesman for the company. In 1921, his hard work was rewarded when he was named president of the corporation. He remained in that capacity until his death in 1936. Rogers's business acumen was not limited to stoves. He built the Hotel Rogers and also developed one of the finest farm complexes in the Midwest. Spring Brook Farms dated to 1842, but it was Rogers's ownership that took the estate to its zenith. The farm was known for breeding high-quality racing and show horses.

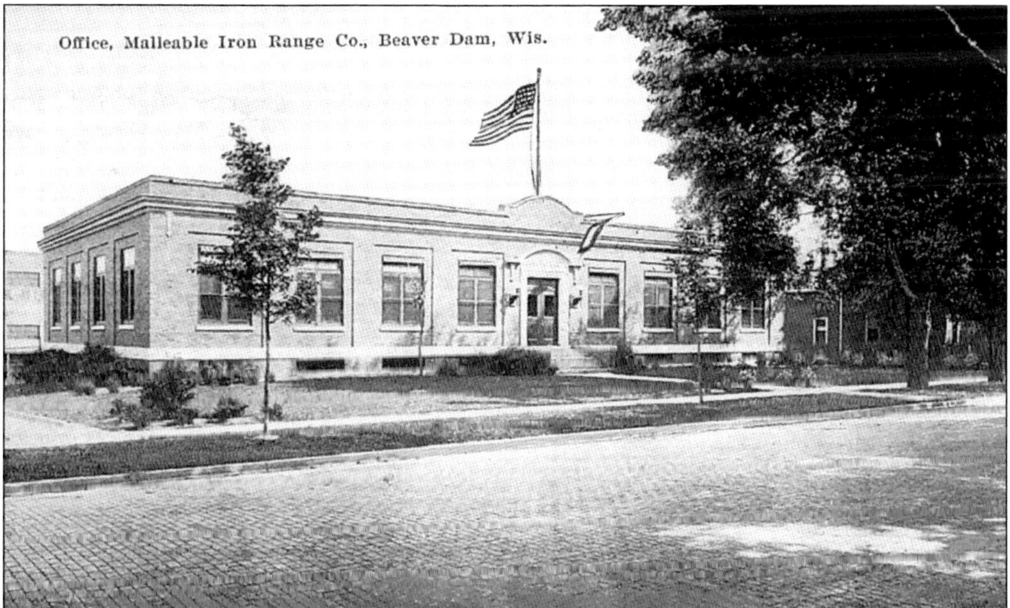

The Malleable Iron Range Company Office. The company initially purchased the facilities of the Beaver Dam Implement Manufacturing Company. In 1916, it erected this office building on North Spring Street.

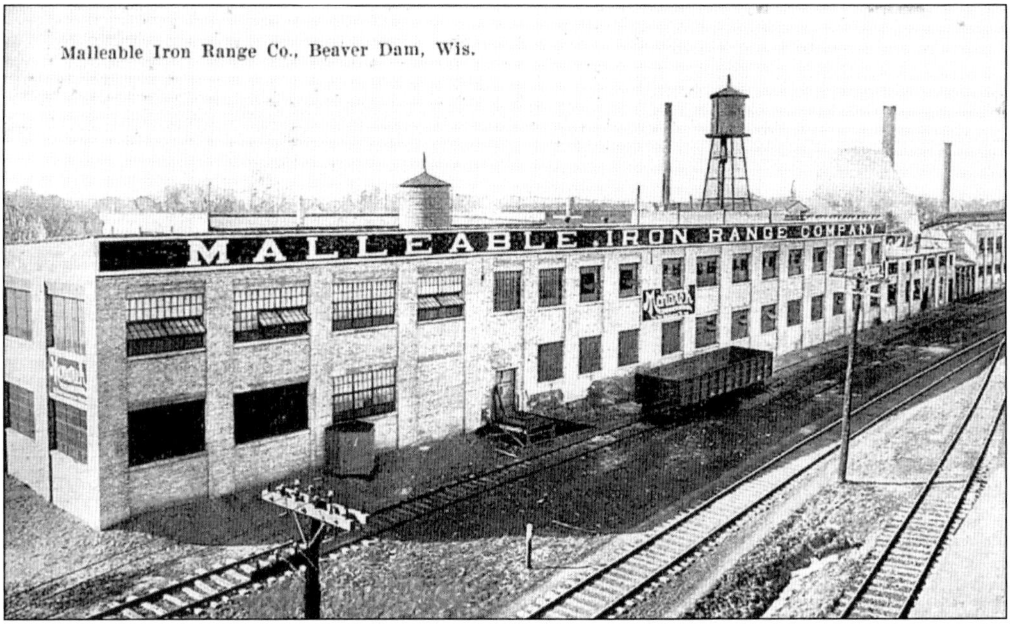

The Malleable Iron Range Company Factory. The Malleable Iron Range Company complex covered a 13.5-acre parcel along North Spring Street. In 1951, the company bought land on Jackson Street to construct an artillery shell plant that would aid in the Korean War effort. Although the company had a successful run for more than three-quarters of a century, by the 1980s, it was having difficulty competing with large national brands. In 1985, the company filed for bankruptcy. The assets were sold, and production wound down. The final blow came in 1990, when the complex was destroyed by fire.

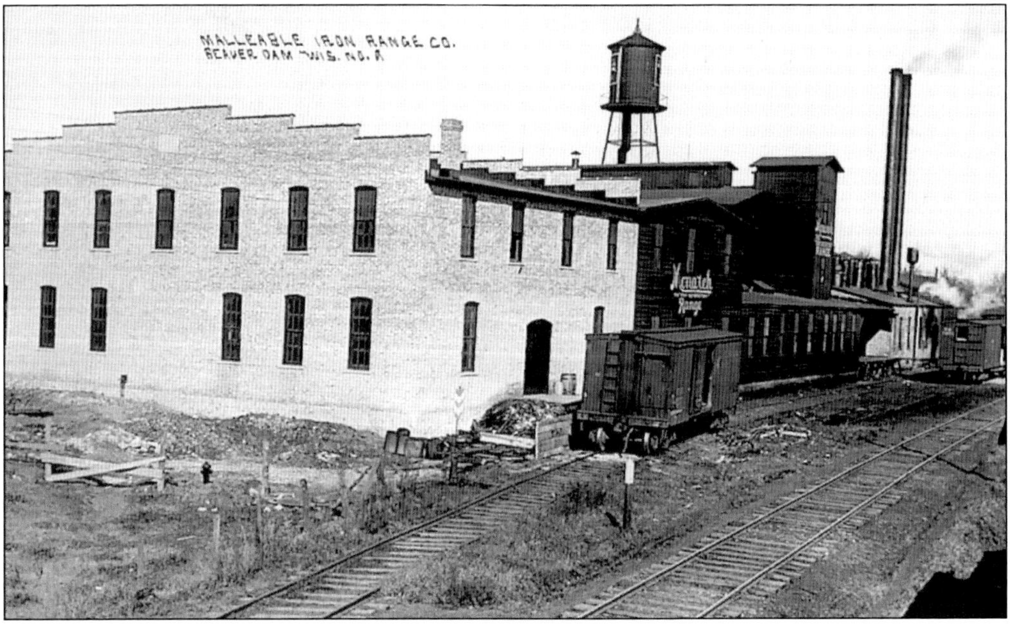

THE OLD WATER TOWER. The city's original water tower was located at the northwest corner of Grove and Norris Streets.

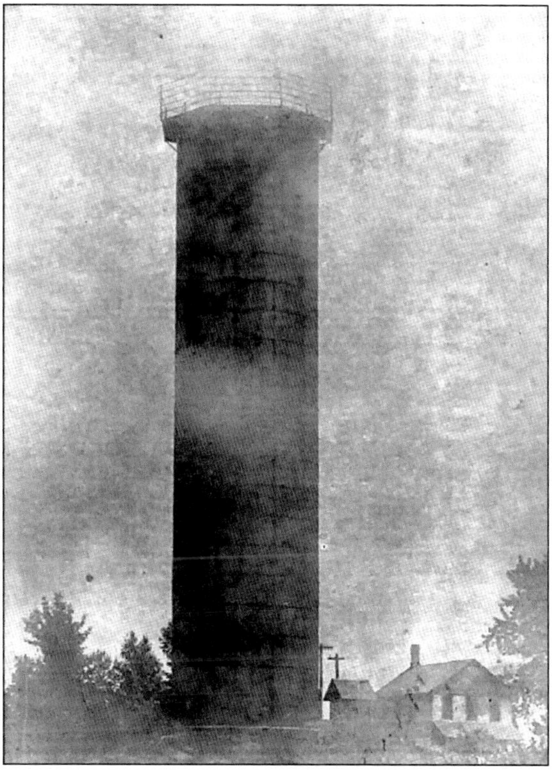

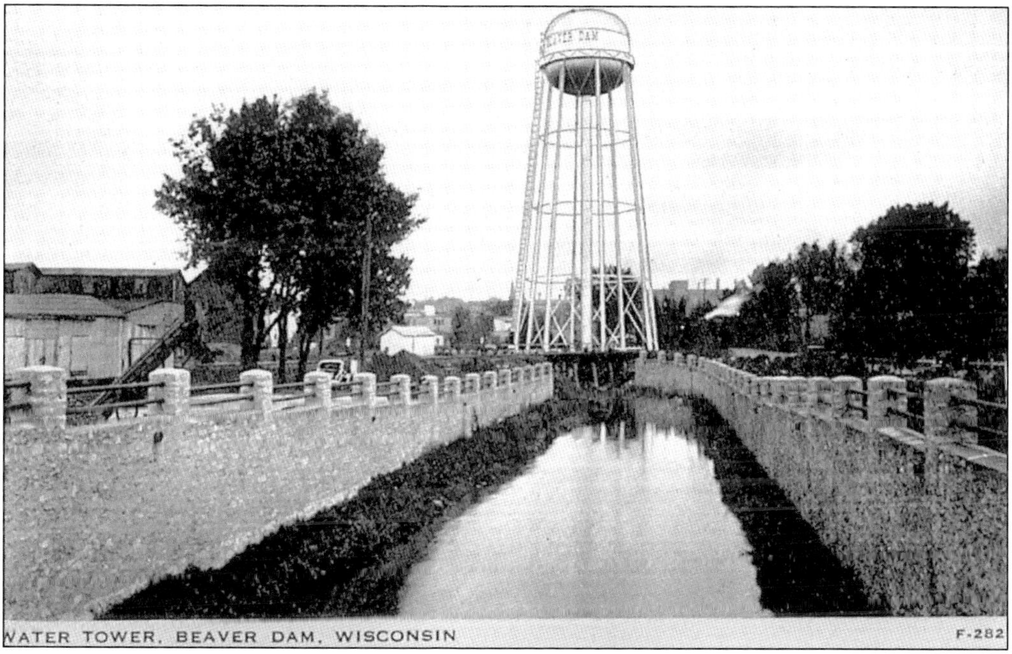

THE NEW WATER TOWER. The old water tower was replaced in 1937 as part of the river plat project. Begun in 1931, the project sought to replace the unsightly area behind the buildings on the south side of Front Street with a sorely needed parking lot. The project ended with construction of the new water tower.

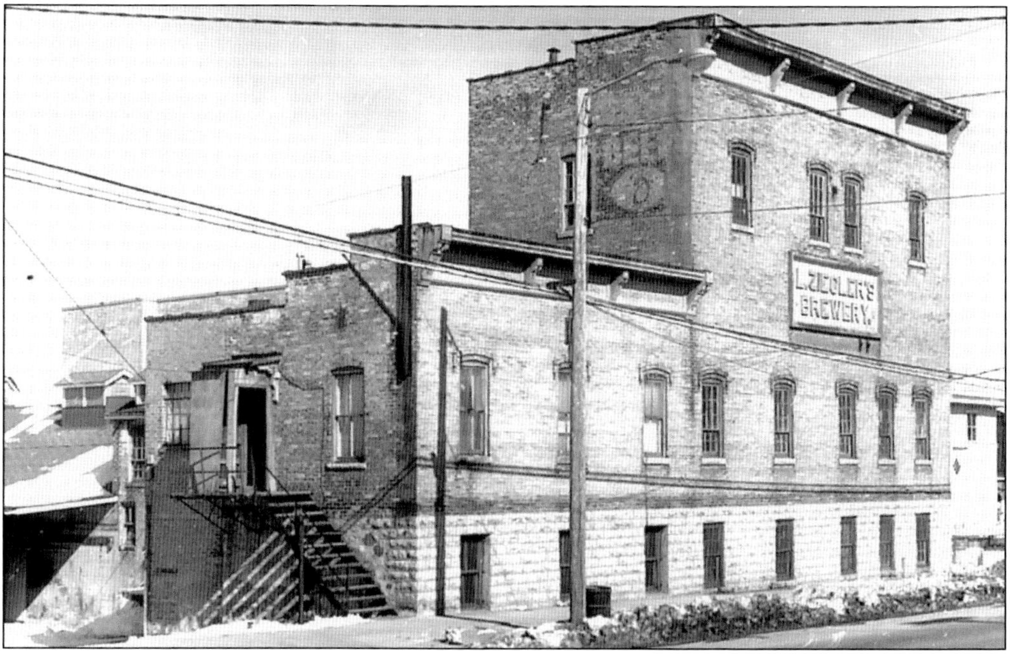

THE ZIEGLER BREWERY. Brewer of the popular Pilot and 5-20 beers, the Louis Ziegler Brewing Company dates from 1905. In that year, Ziegler purchased the city's first brewery, originally built by Michael Biersack on Madison Street. Ziegler constructed a new brewery on the grounds in 1906. The business stayed in the family until 1949. Four years later, the facility closed.

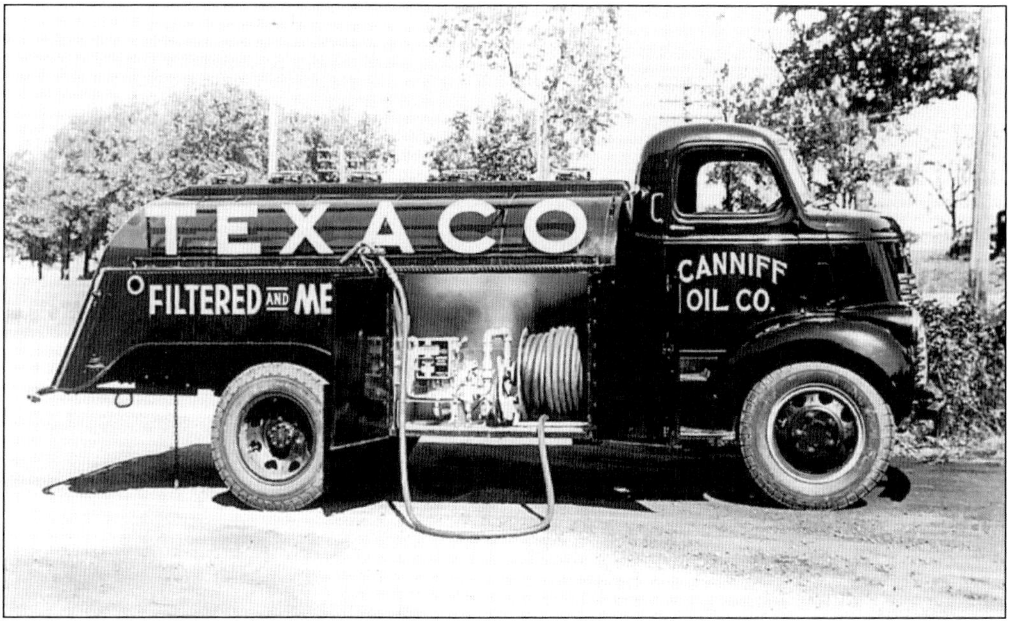

THE CANNIFF OIL COMPANY. Located at 1307 North Spring Street, the Canniff Oil Company was a distributor of Texaco oil and Goodrich products.

Six

All I Want Is Loving You and Music, Music, Music

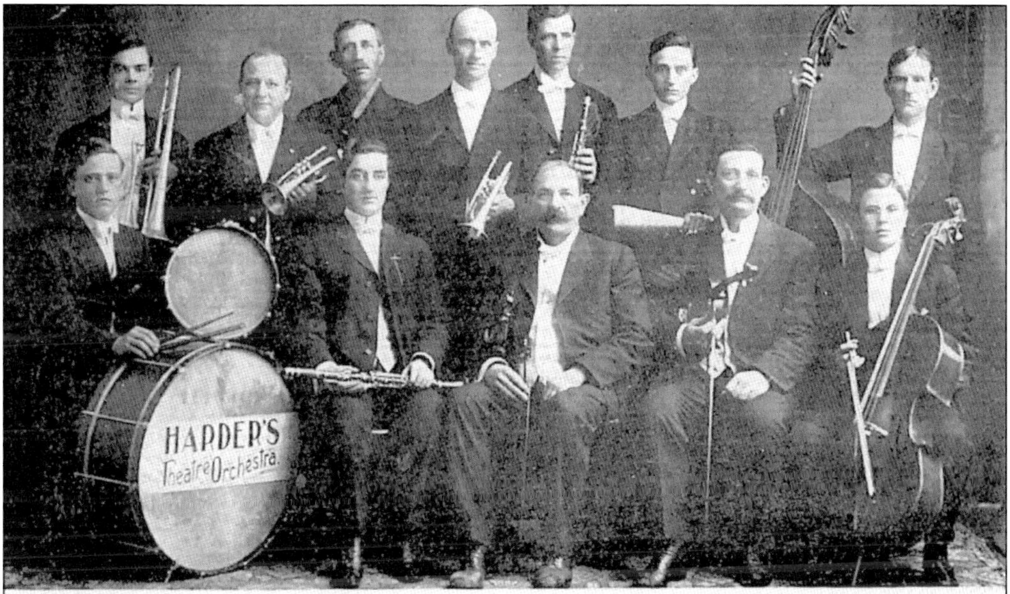

Harder's Theatre Orchestra. During the late 19th and early 20th centuries, the Harder family played a significant role in the musical life of Beaver Dam. Harder's Military Band performed in the Vita Park Bandstand, in the Vita Hotel, and at the grand balls of the day. Harder's Theatre Orchestra played for local community theaters and manned the pits to accompany silent movies at the Odeon and Davison Theatres. Pictured, from left to right, are the following: (first row) Frank Henninger, Judge Roedl, director John Harder, Stan Yasger Sr., and Otto Ostensacken; (second row) Otto "Windy" Jacobs, L. Henninger, Ed Hohl, Ed Holleck, John Schlitzberg, John Yasger, and William Rupple.

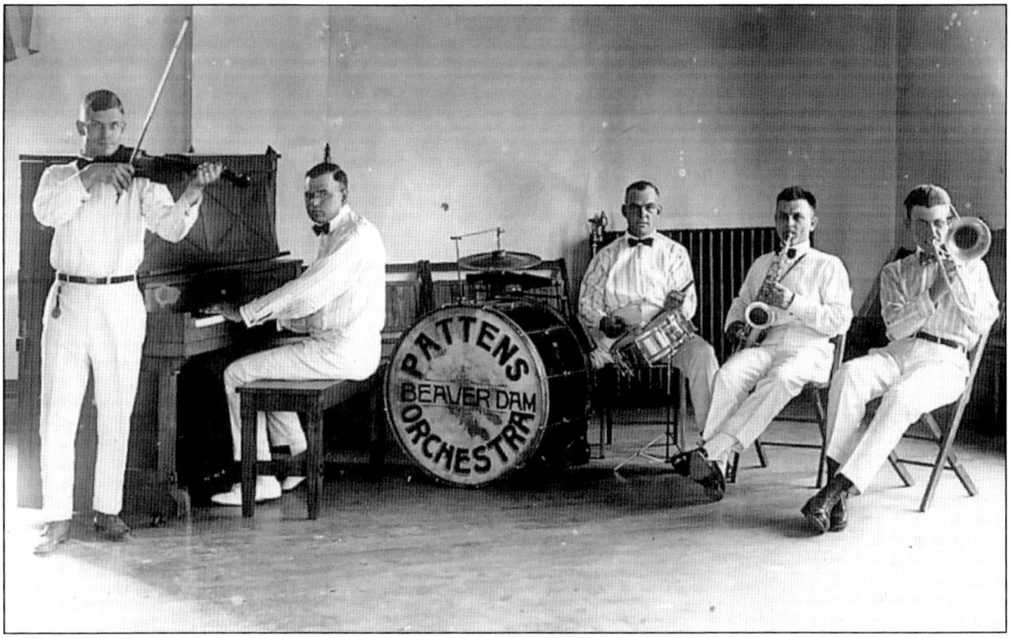

PATTENS ORCHESTRA. One of the popular local bands of the 1930s was Pattens. Pictured are, from left to right, Ira Patten, Reuben Hachenberger, Del Freeman, Lerum Luck, and Elmore Schiller.

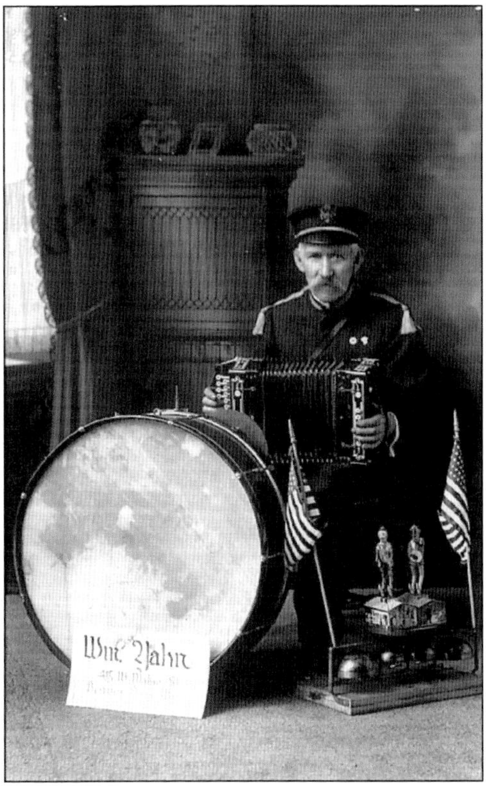

ONE-MAN BAND WILLIAM YAHN. William Yahn was one of several one-man bands that entertained Beaver Dam audiences. He worked out of his home at 415 West Mackie Street.

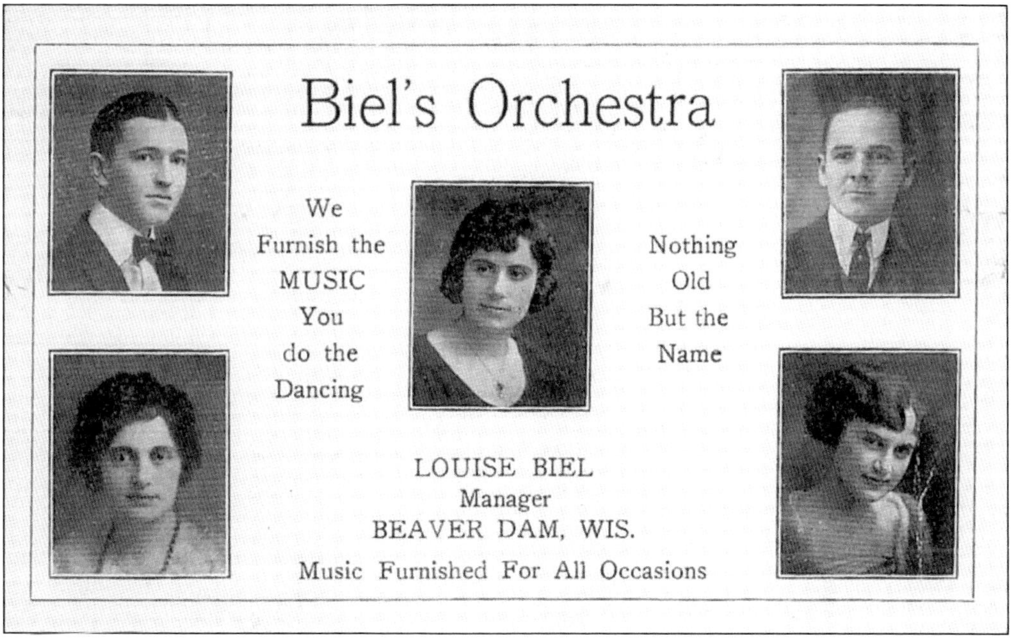

BIEL'S ORCHESTRA. Originally from Lost Lake, the Biel Family Orchestra and the Biel Girls Orchestra delighted area audiences from the second decade of the 20th century to the 1920s. During its many incarnations, members of the Biel orchestras included Jennie Biel Shesky, Henrietta Biel Smith, Laura Biel, Louis Biel, Agnes Biel Garvin, Ed Biel, and Emil Biel.

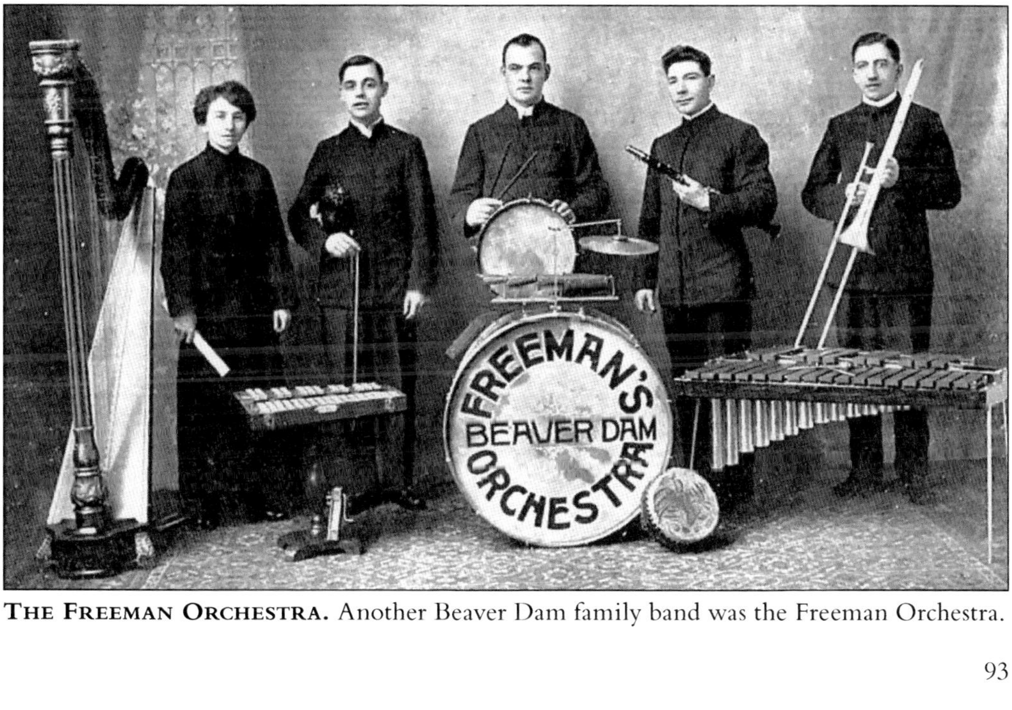

THE FREEMAN ORCHESTRA. Another Beaver Dam family band was the Freeman Orchestra.

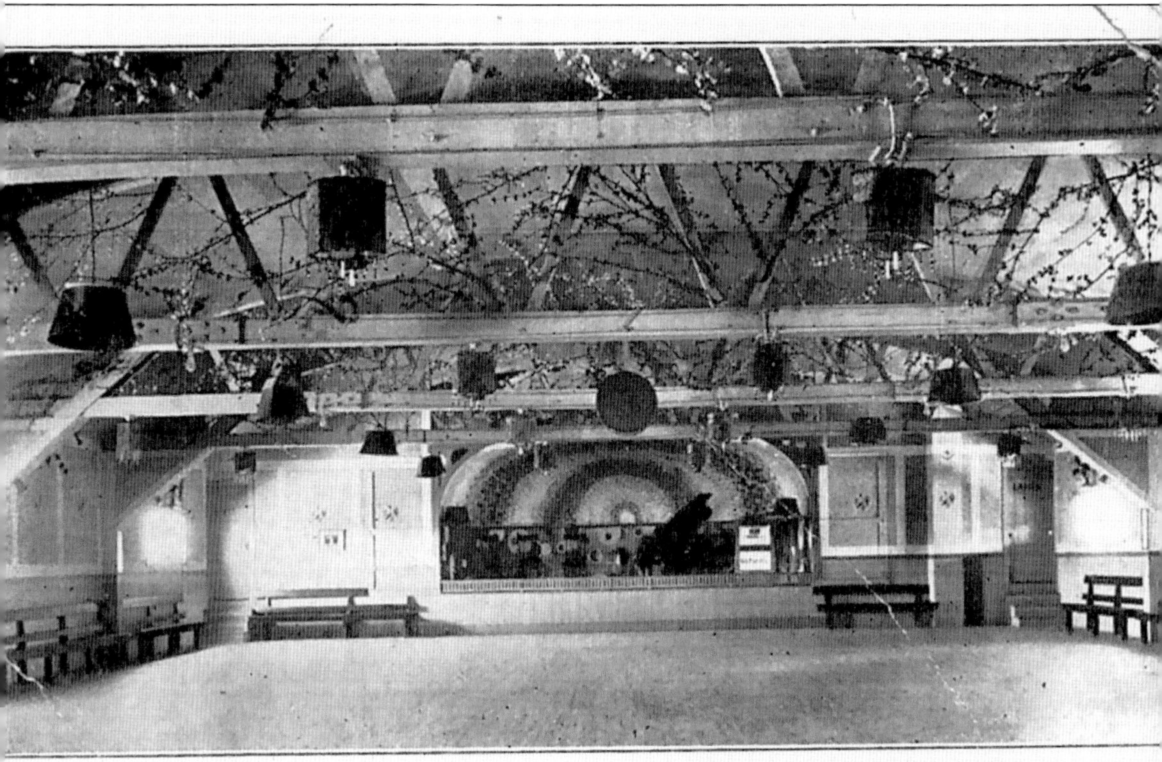

Inside View of the Dance Hall at Crystal Lake Beach, Beaver Dam, Wis.

THE CRYSTAL LAKE BEACH DANCE HALL. One of the many establishments that hosted local musical talent was the Crystal Lake Beach Dance Hall. The hall, managed by Louis Payne, doubled as a roller-skating rink.

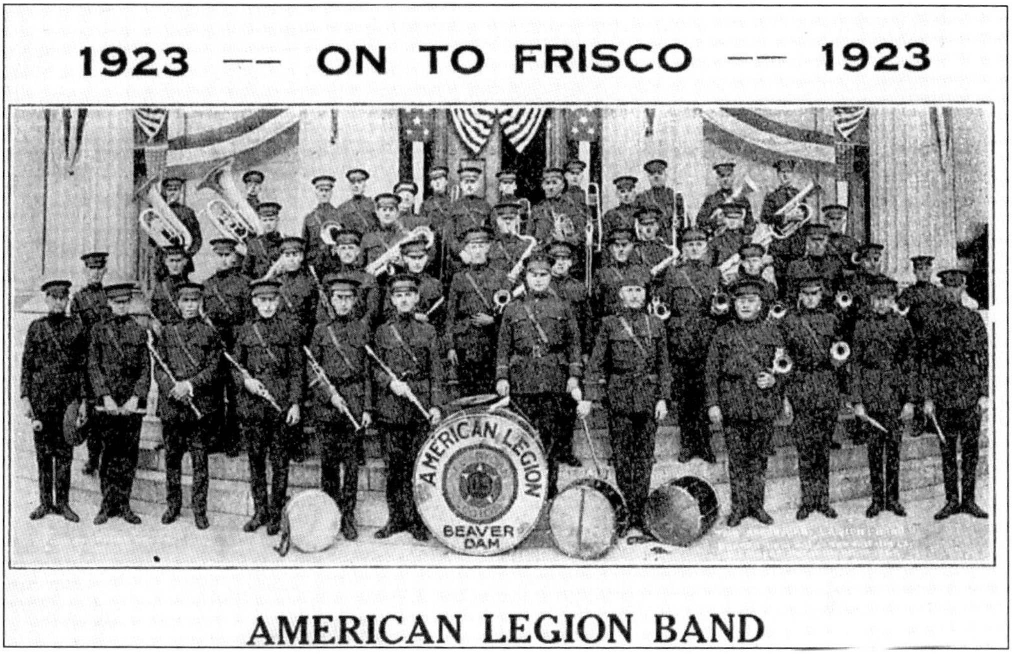

THE AMERICAN LEGION BAND, 1923. Taking over the duties as city band after Harder's Band dissolved, a group of former doughboys organized the American Legion Band in 1921. The band entered American Legion competitions throughout the United States, Canada, and Mexico. The group disbanded in 1955.

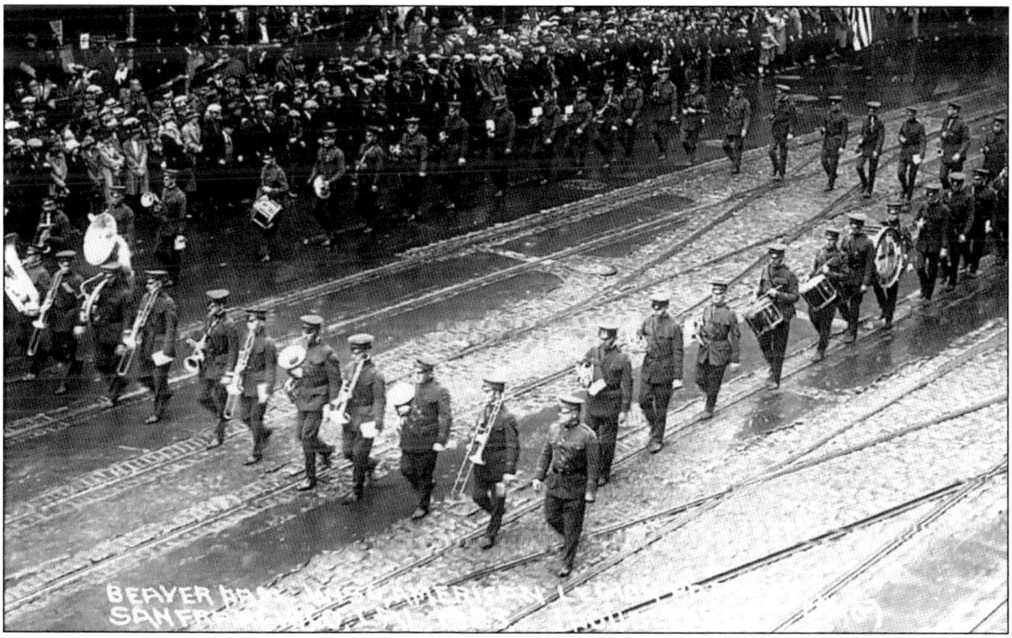

A BAND TRIP TO SAN FRANCISCO, 1923. The American Legion Band entered its second competition in San Francisco in 1923.

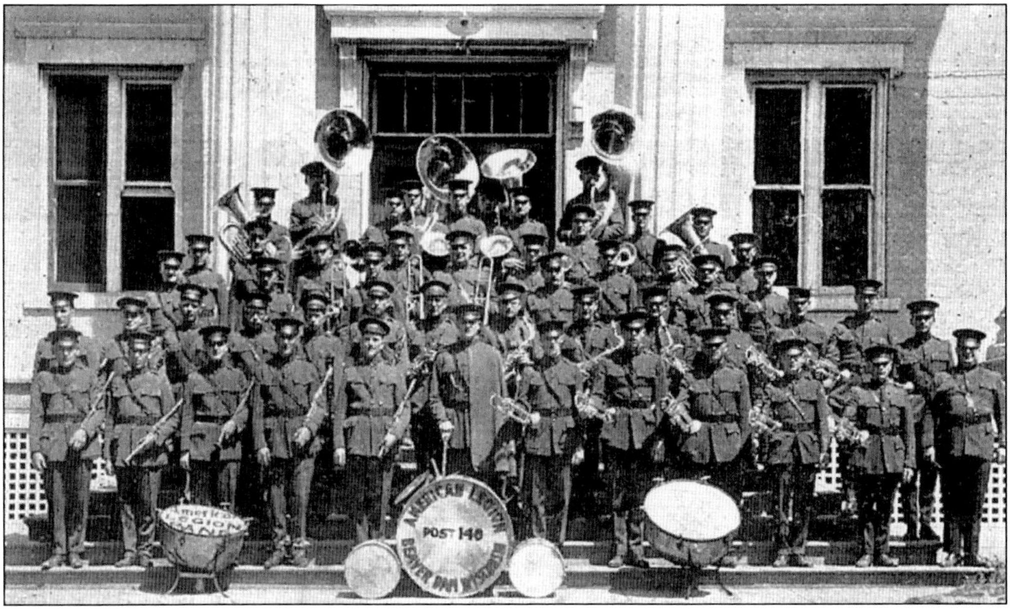

The American Legion Band, 1924. In 1924, the band journeyed to St. Paul, where they placed in the competition. The tall horn player in the center of the back row is future movie star Fred MacMurray.

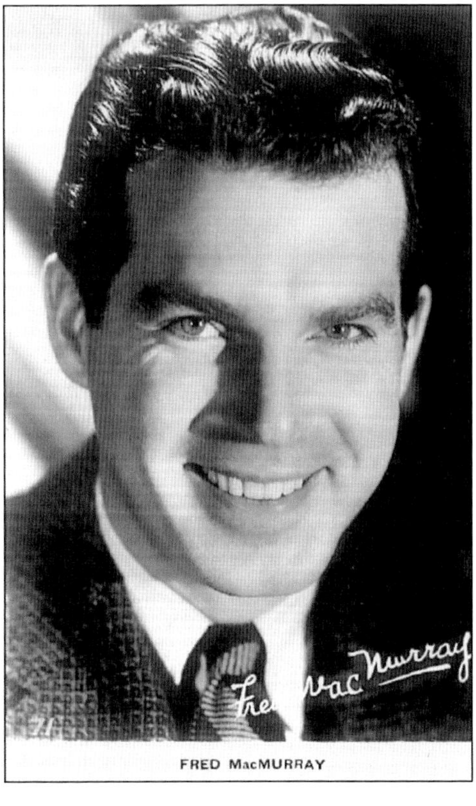

Movie Star Fred MacMurray. Fred MacMurray was one of Beaver Dam's most famous musicians. Few realize that his career was based upon his musical talents. Brought to Hollywood in 1934 as a Broadway musical performer, MacMurray sang or played an instrument in 15 of his 80-plus movies. MacMurray was the subject of many a postcard. This postcard was printed in Great Britain in 1936.

Seven

SWAN CITY: HOW I LOVE YA, HOW I LOVE YA

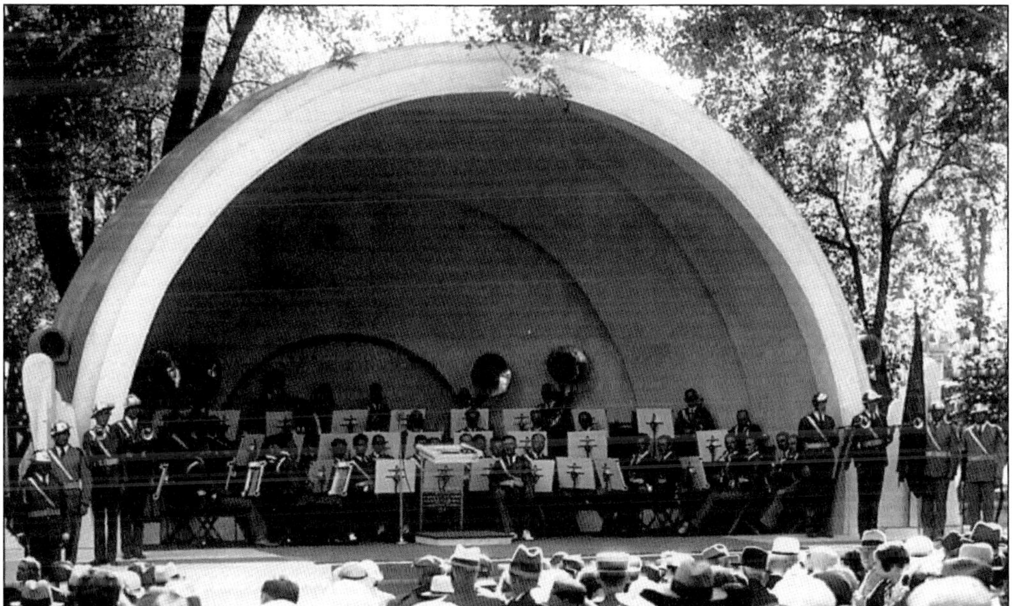

THE SWAN CITY PARK BANDSHELL. The bandshell, erected in 1935, was modeled after the Swift and Ford bandshells of the Chicago Century of Progress. At evening shows, crowds were entertained by the dancing waterfront, lit by 780 colored lights.

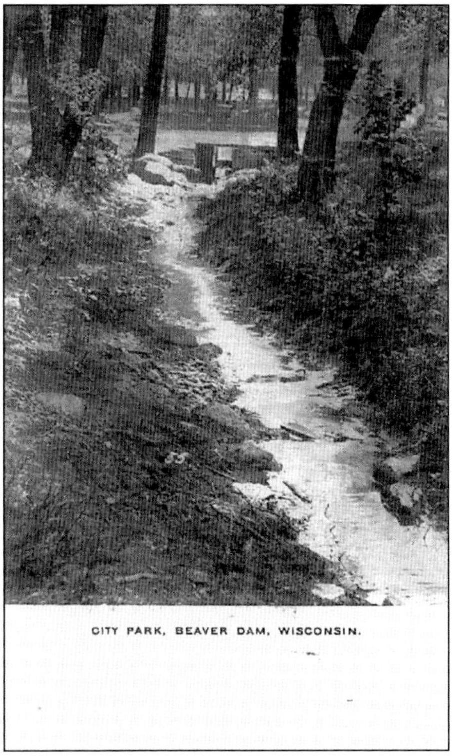

CITY PARK, C. 1910. In 1905, Vita Park was sold to the city and renamed simply City Park. Over the next 10 years, the park's upkeep was neglected. The shrubbery and lawns were permitted to overgrow, and the wood bridges deteriorated into dangerous passageways.

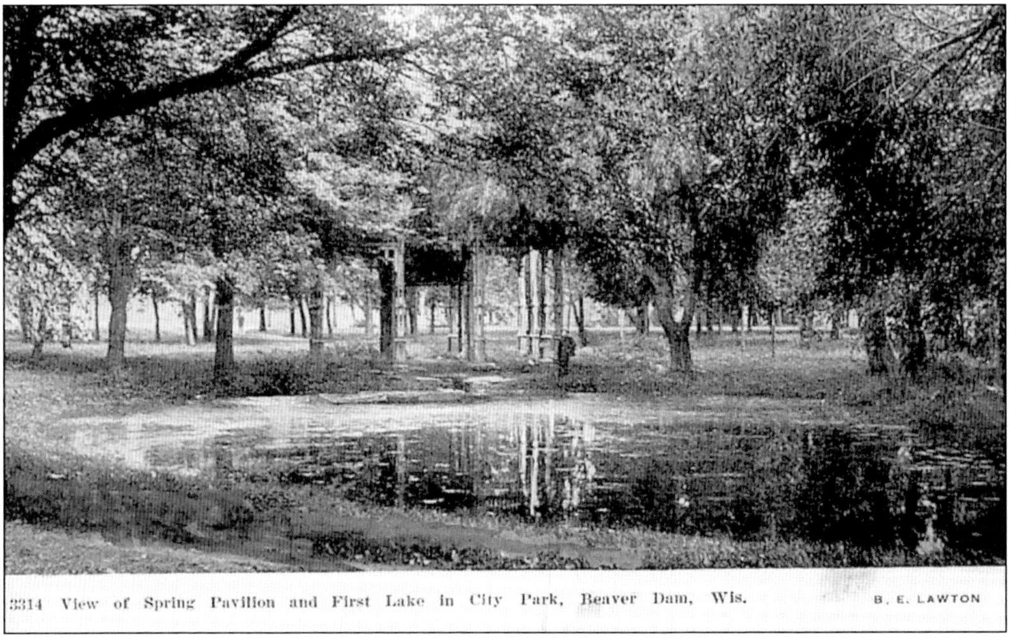

THE SPRING PAVILION, C. 1910. As seen in this photograph, the once majestic pavilion was in constant disrepair, and the spring was sealed over.

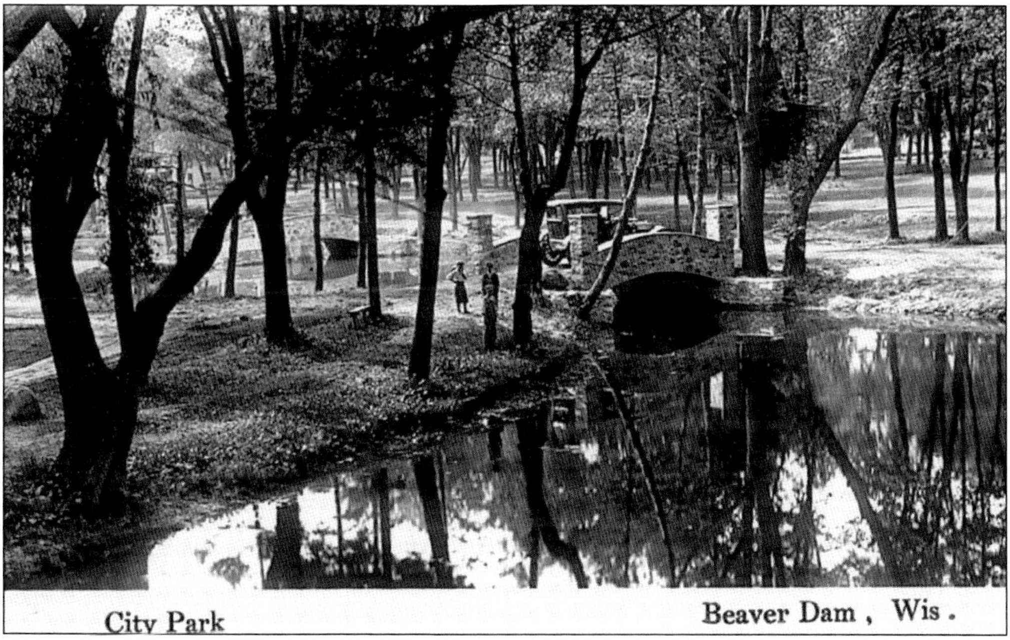

The City Park Bridge, 1915. In 1915, a challenge to city citizens drew an army of volunteers, who returned the park to its former splendor. Lawns were trimmed, new shrubbery was planted, pathways were repaired, and new stone bridges were constructed. The public soon returned in droves to enjoy the improvements.

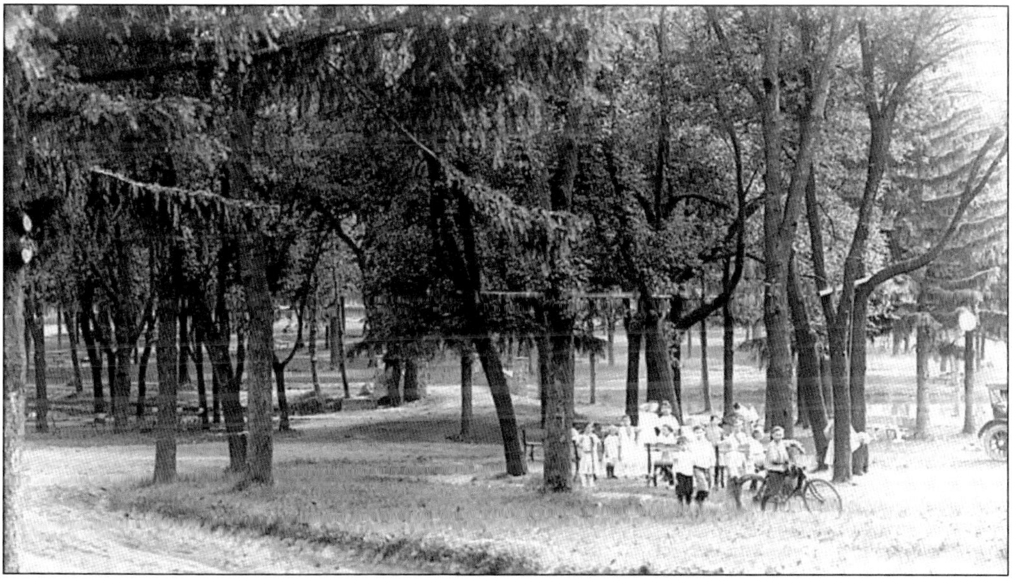

A City Park Party. Once again, the park became a place of pride for the city, providing a popular place for children's gatherings. City Park was renamed Swan City Park in honor of its founder.

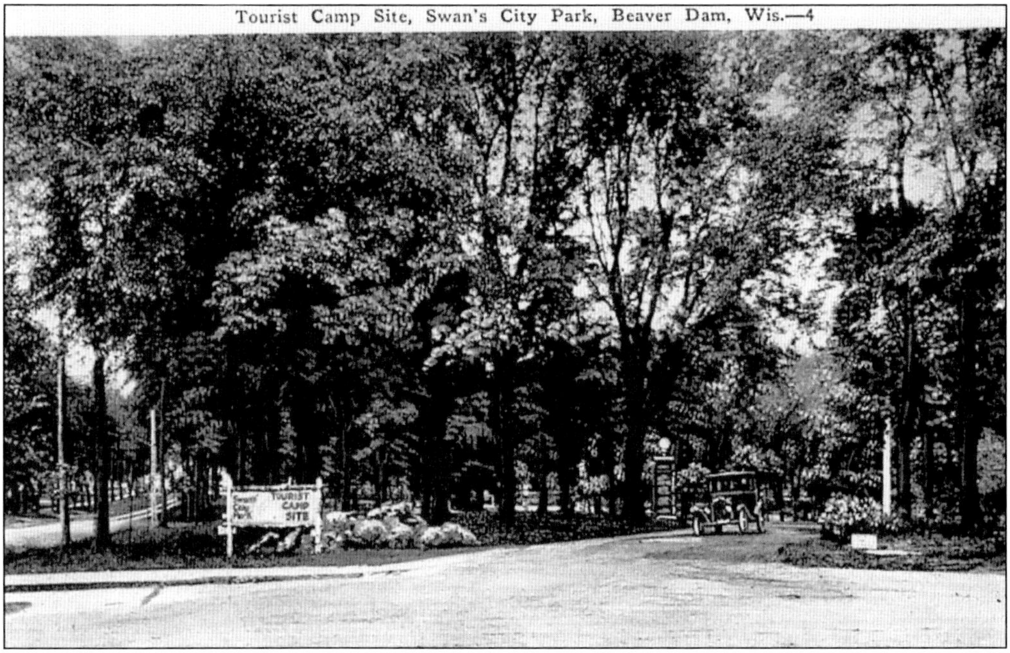

THE SWAN CITY PARK TOURIST CAMPSITE, 1920S. In the 1920s, the city offered visitors the added attraction of camping in the park.

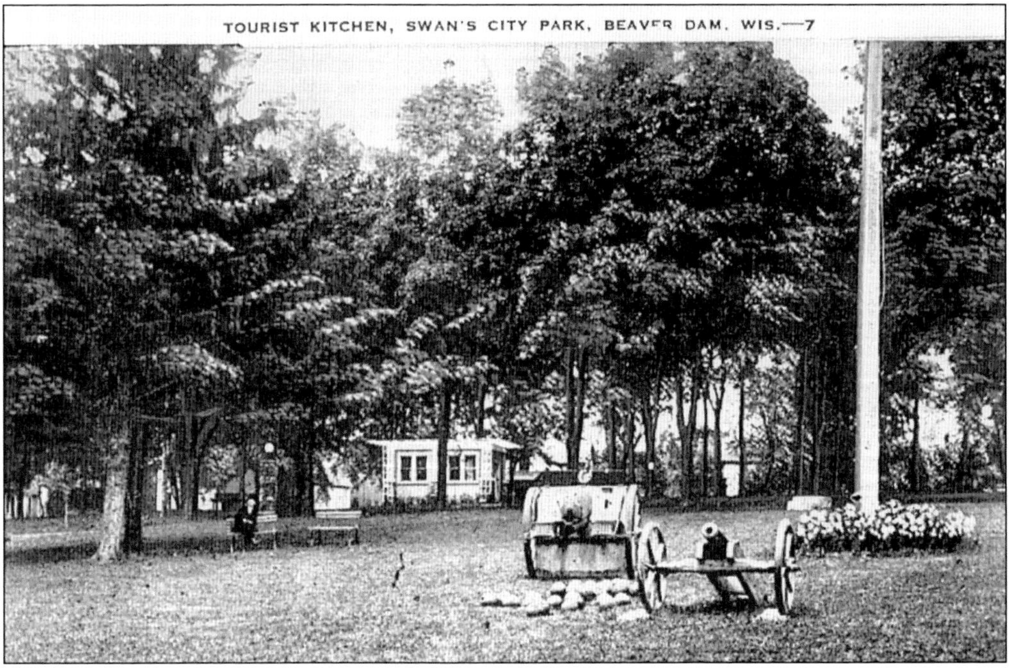

THE TOURIST KITCHEN. A kitchen was built to accommodate the needs of both campers and day visitors.

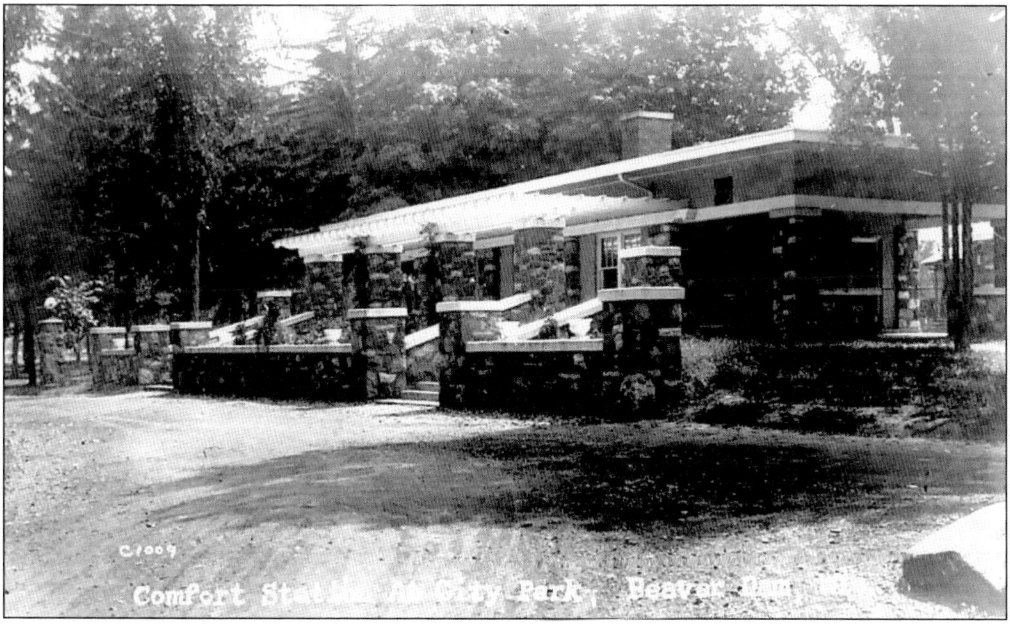

THE COMFORT STATION. The comfort station provided state-of-the-art bathing facilities for campers. The tourist campsite closed in 1937, when the severe effects of the Great Depression spurred residents to pitch permanent tents on the grounds.

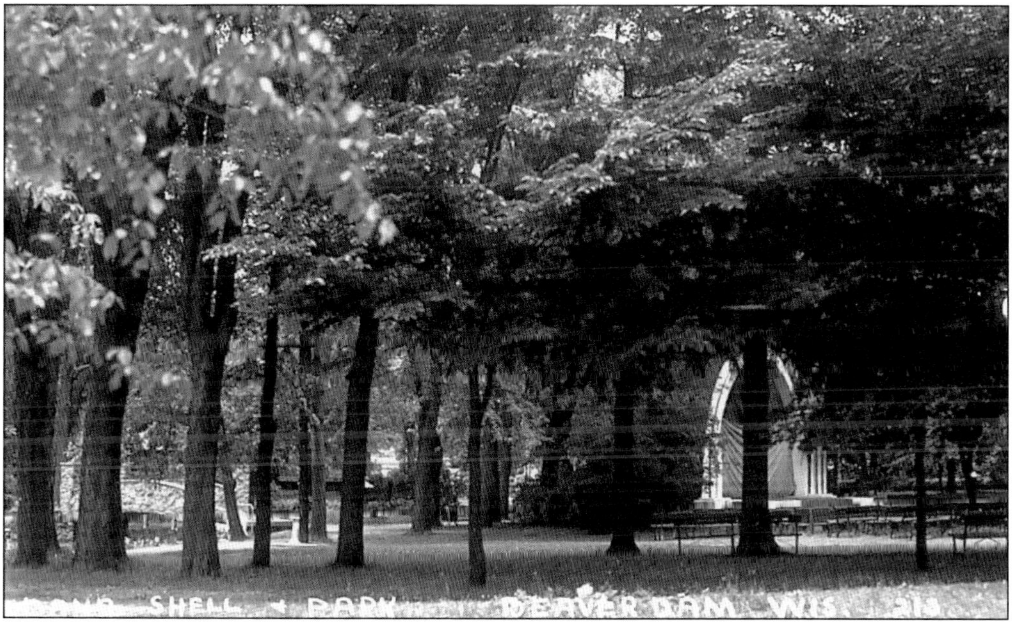

THE BANDSHELL. The park offered weekly concerts at the bandshell. Although the city had provided seating for 3,000 attendees, the concerts proved so popular that attendance regularly topped 10,000. That translates to 7 out of every 10 city residents present in the park on concert evenings.

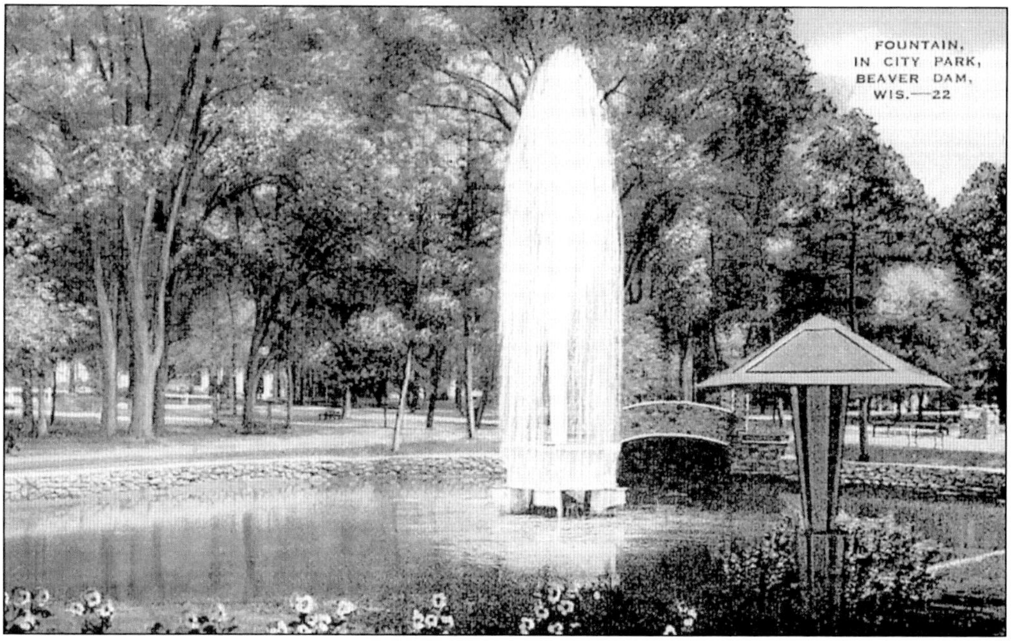

THE SINGING FOUNTAIN. In 1938, the park added a singing fountain in the second lagoon. The fountain possessed numerous spray and color combinations as music was piped through two large speakers below. This fountain and the dancing waterfront were the brainchild of Orrin Hofferbert, an engineer at the Malleable Iron Range Company.

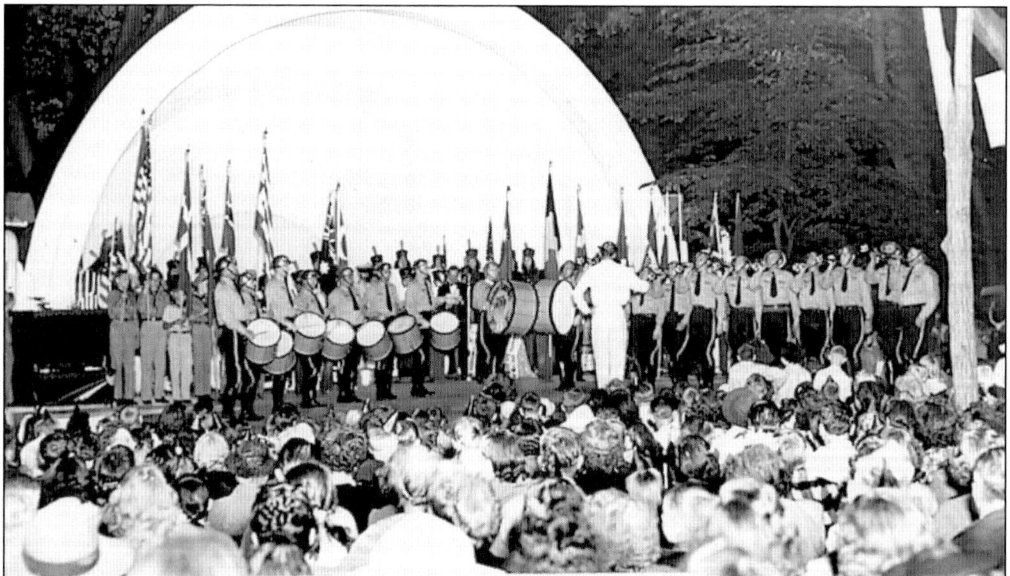

CIRCUS NIGHT. From the 1940s through the 1960s, Swan City Park presented two unique concerts at the bandshell each year. Circus Night brought in professional circus acts as well as local talent. On Indian Night, Native Americans from Wisconsin Dells performed tribal dances, and Mrs. Marvin Keil sang her unforgettable rendition of "Indian Love Call."

Eight

Get Me to the Church on Time

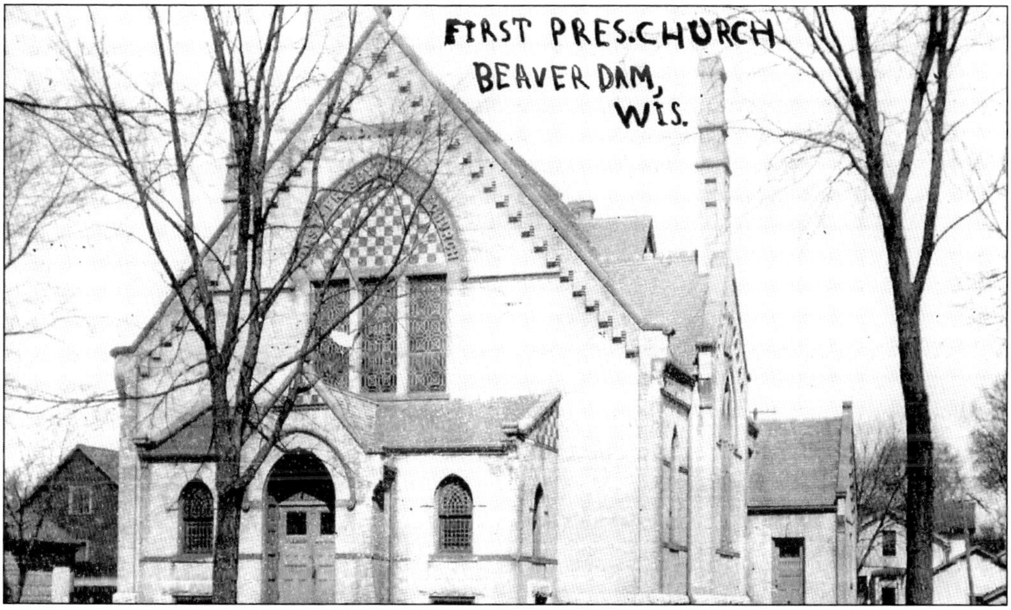

The Grace Presbyterian Church. The Grace Presbyterian Church dates back to the founding fathers of Beaver Dam. The congregation held its first service in 1843 at the Thomas Mackie home on Front Street. After constructing three smaller churches, the congregation built this ornate brick church at 116 Park Avenue in 1889. The M&I Bank now occupies this site.

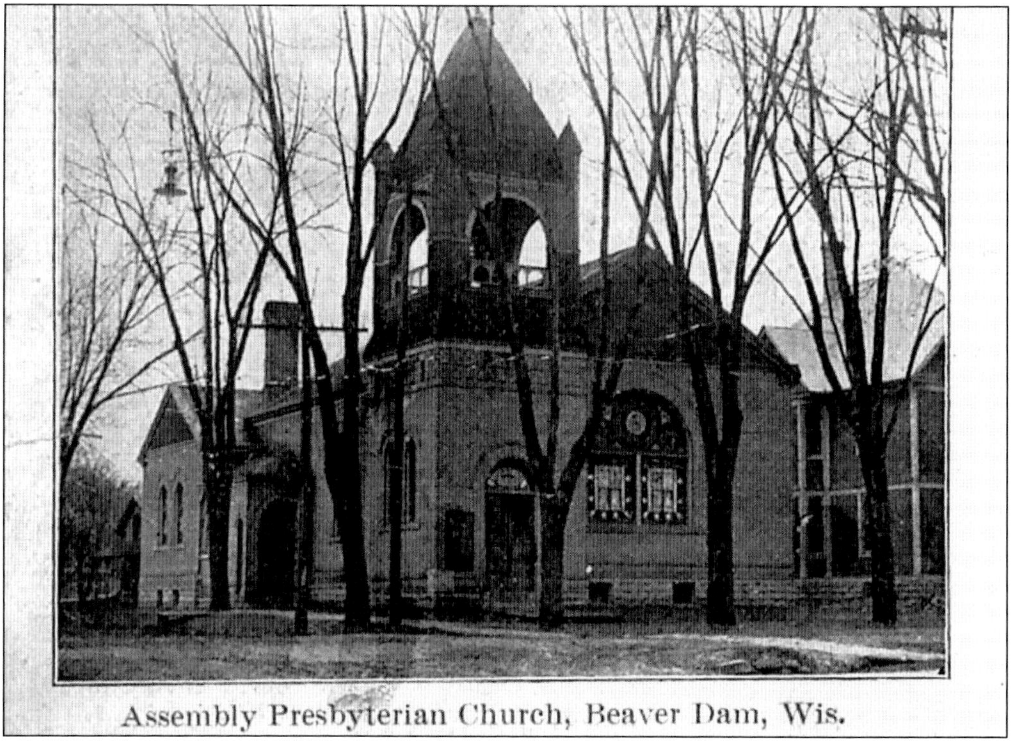

THE ASSEMBLY PRESBYTERIAN CHURCH. The Assembly Presbyterian Church was organized in Beaver Dam in 1858. In 1859, this church was erected at the corner of Lincoln Street and East Maple Avenue. The church was bricked and remodeled in 1900. In 1929, the congregation dissolved and merged with the Grace Presbyterian Church.

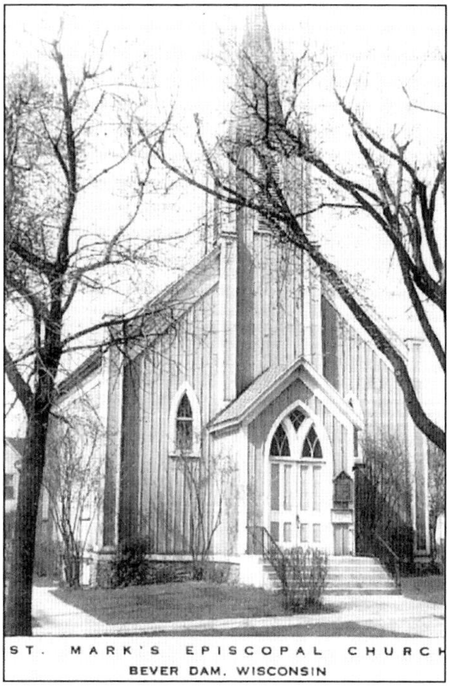

ST. MARK'S EPISCOPAL CHURCH. St. Mark's Episcopal Church was established in 1855 as a mission church at the corner of Beaver and Front Streets. Three years later, the church building pictured here was moved to the corner of Maple Avenue and Lincoln Street. The building was sold in 1978 and is currently being used as a preschool. St. Mark's purchased the Assembly of God Church on East Mill Street. The congregation celebrated its 150th anniversary in 2005.

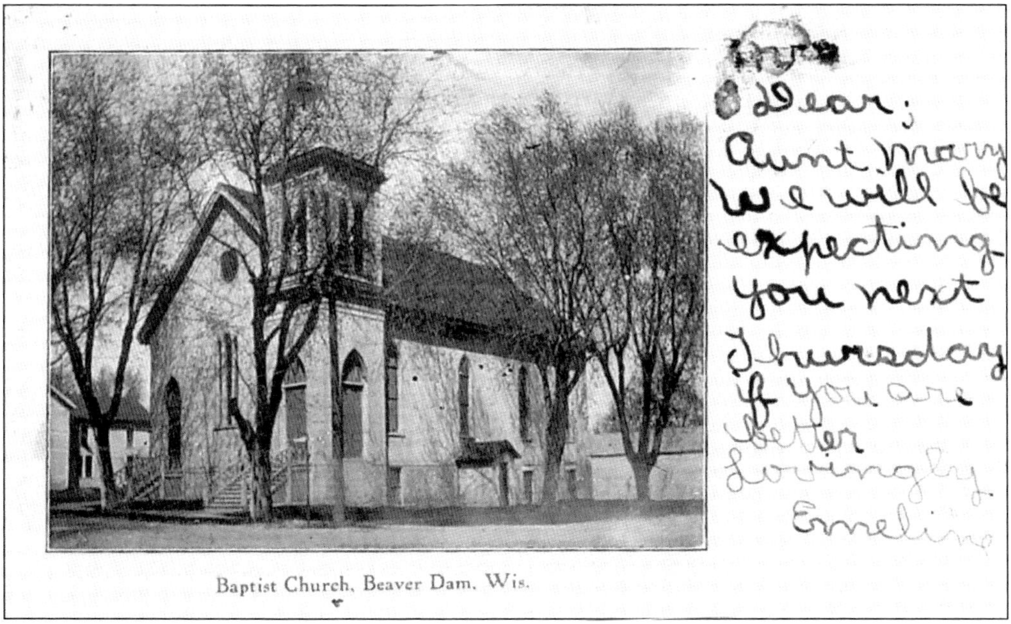

THE OLD BAPTIST CHURCH. This building, located at the corner of East Maple Avenue and North Spring Street, was built in 1855. In 1865, the church was destroyed by fire. It was rebuilt, but it succumbed to fire again in 1911.

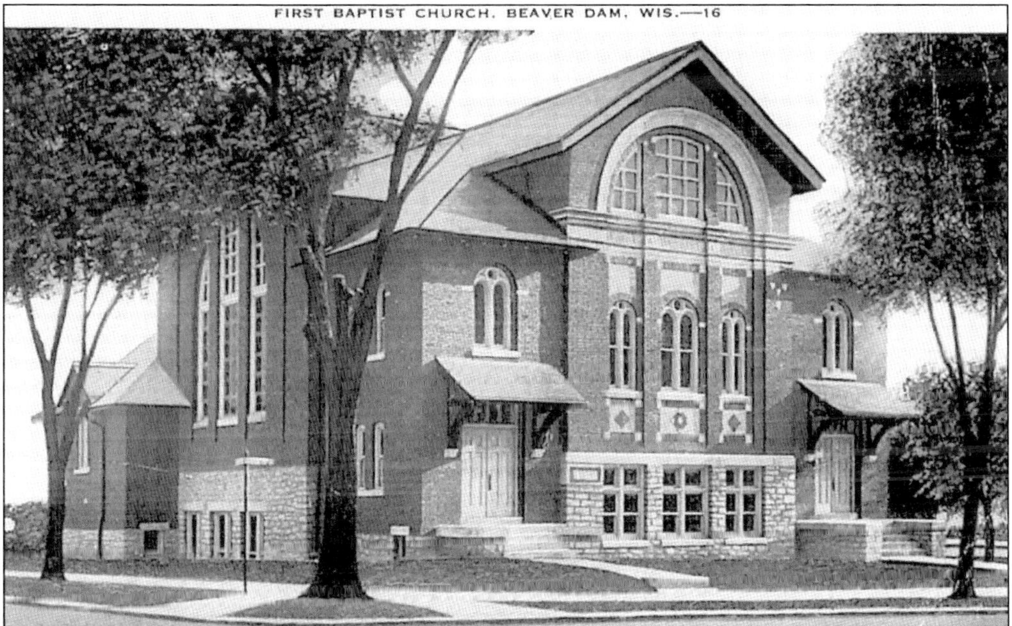

THE NEW BAPTIST CHURCH. After the fire of 1911, the Baptist congregation moved one block north. The Baptists remained at this site until 1972, when they sold the building to the Beaver Dam Community Theatre.

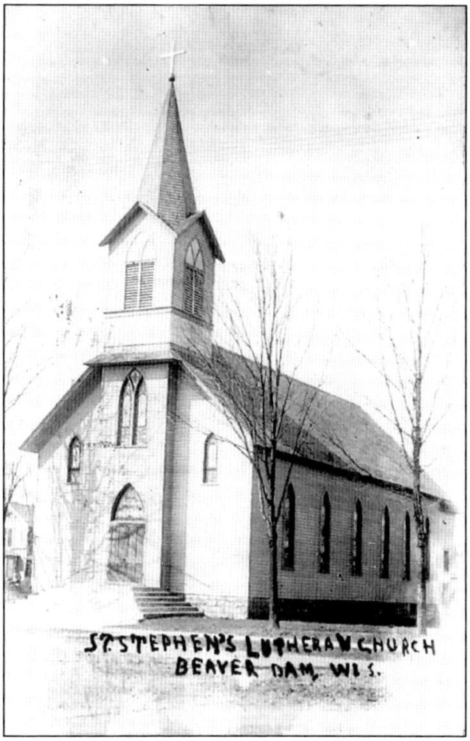

THE OLD ST. STEPHEN'S LUTHERAN CHURCH. St. Stephen's Evangelical Lutheran Church was established in 1875. The original church was located at the corner of Rowell and South Center Streets. In 1883, the church was moved across the Madison Street Bridge to the congregation's new location at West Maple Avenue and West Street.

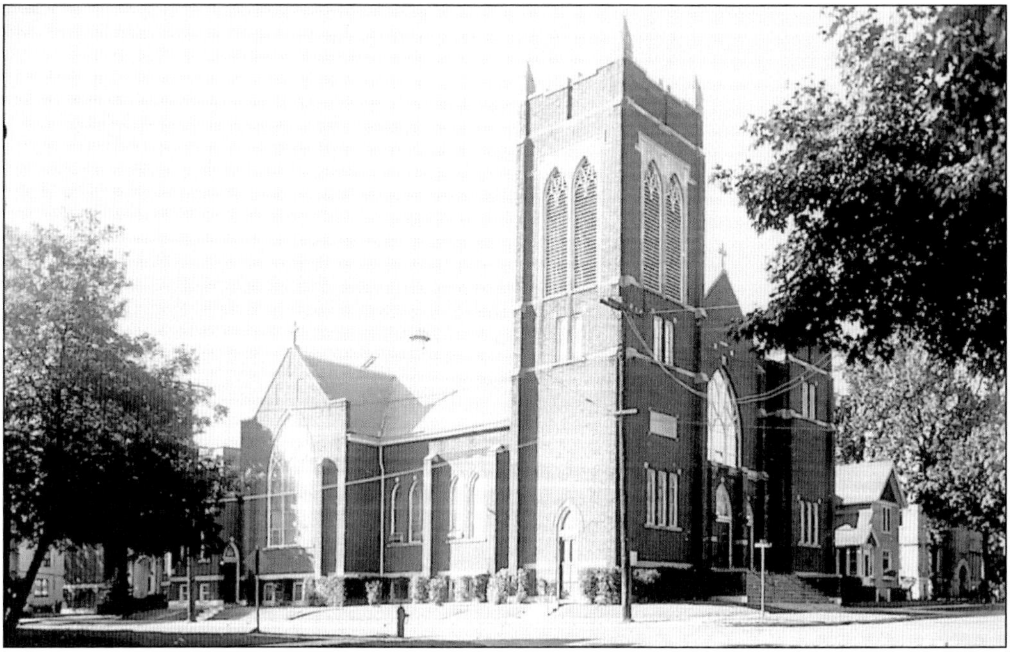

THE NEW ST. STEPHEN'S LUTHERAN CHURCH. The new St. Stephen's Lutheran Church was built in 1919 and is still in operation today. The old church was placed alongside the new church and utilized as a parochial grade school.

ST. JOHN'S GERMAN LUTHERAN CHURCH. Established in 1860, the First Evangelical Lutheran Church, known for a brief time as St. John's German Lutheran Church, was originally located on James Street. In 1892, the southwest corner of Mackie and Center Streets was purchased as the location for a new church. In 1893, the church pictured here was dedicated. Recently, this building was razed, and a new church was placed at the site.

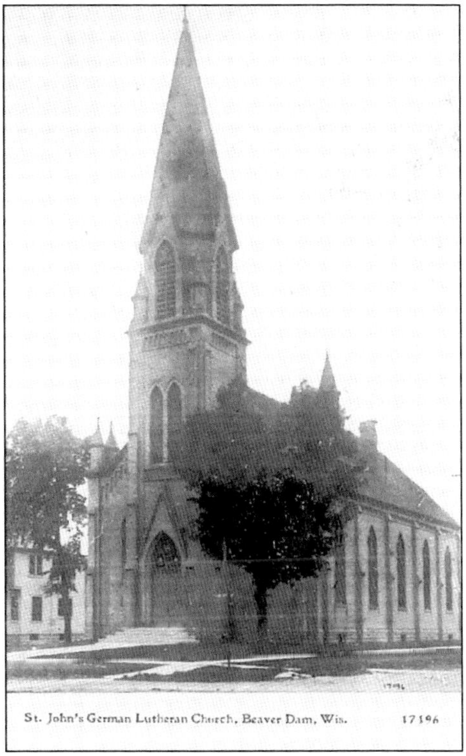

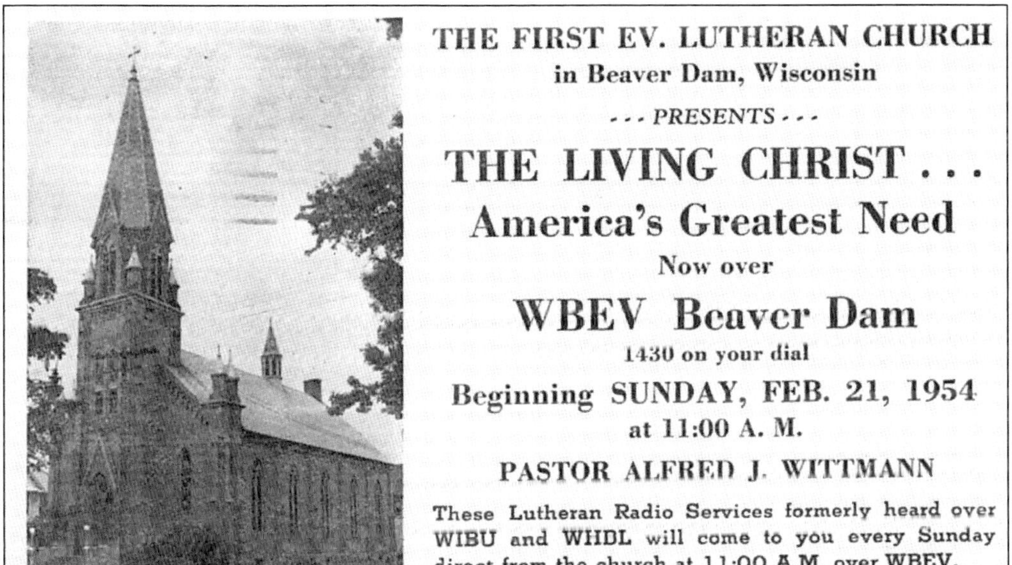

A WBEV ADVERTISEMENT. As early as 1938, the First Evangelical Church broadcast its services over the airwaves on WIBU in Sheboygan and WWCF-FM in Baraboo. In 1954, WBEV of Beaver Dam took over the Sunday service program.

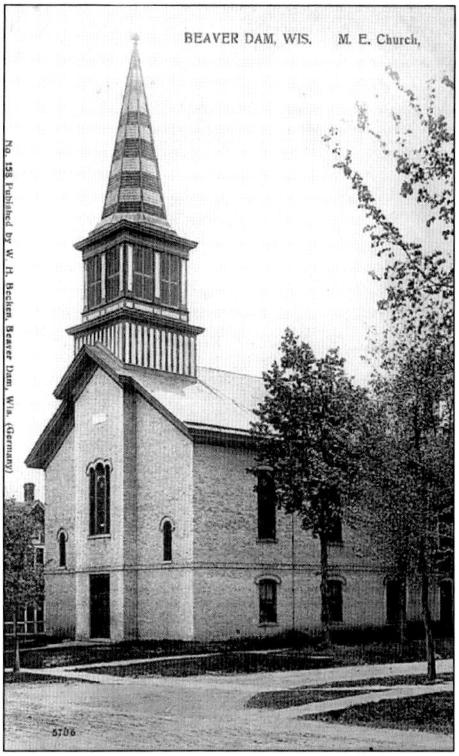

The Methodist Episcopal Church. The Methodist Episcopal Church began services in 1847 at a site located at the corner of Beaver Street and West Maple Avenue. In 1872, a new church, pictured here, was constructed at the corner of North Spring and West Third Streets. In 1962, the church was razed and replaced by the Wisconsin Telephone building.

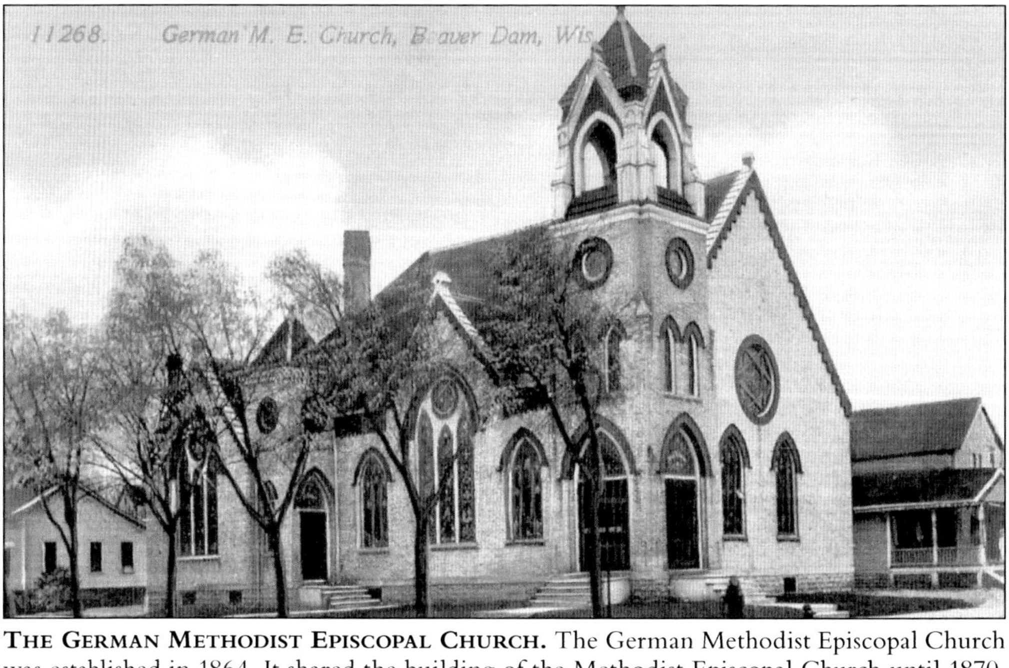

The German Methodist Episcopal Church. The German Methodist Episcopal Church was established in 1864. It shared the building of the Methodist Episcopal Church until 1870. The German Methodist Episcopal Church, pictured here, was established at the corner of Washington and South Spring Streets, down the street from the depot.

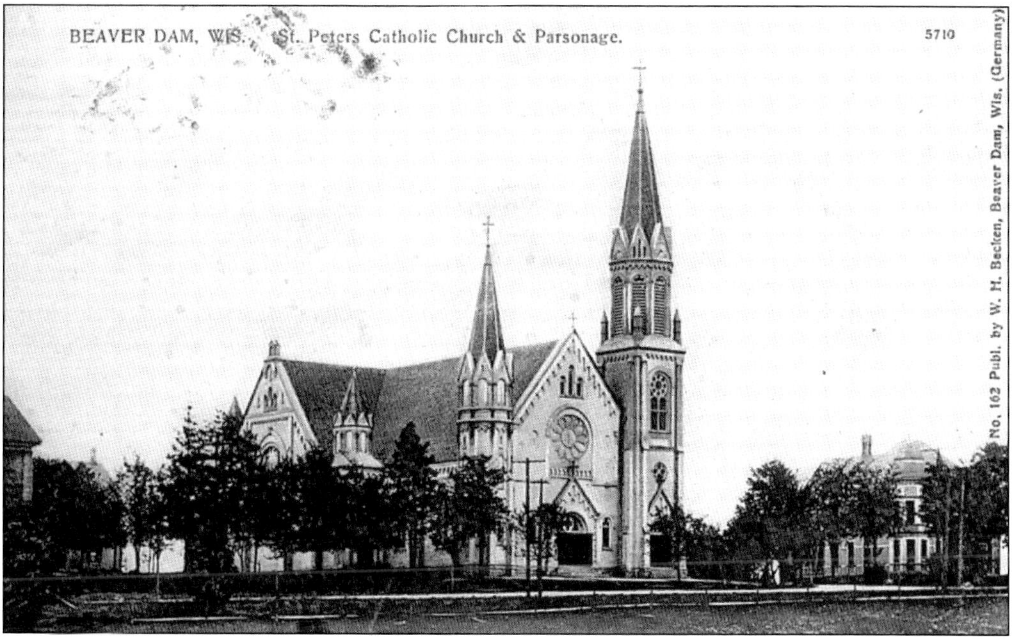

ST. PETER'S CATHOLIC CHURCH AND PARSONAGE. St. Peter's Catholic Church, also known as the German Catholic Church, was the first Catholic church in the city. The first church building was constructed in 1854. This church, the third, was erected in 1900. The parsonage was constructed in 1890, and the convent was built in 1924.

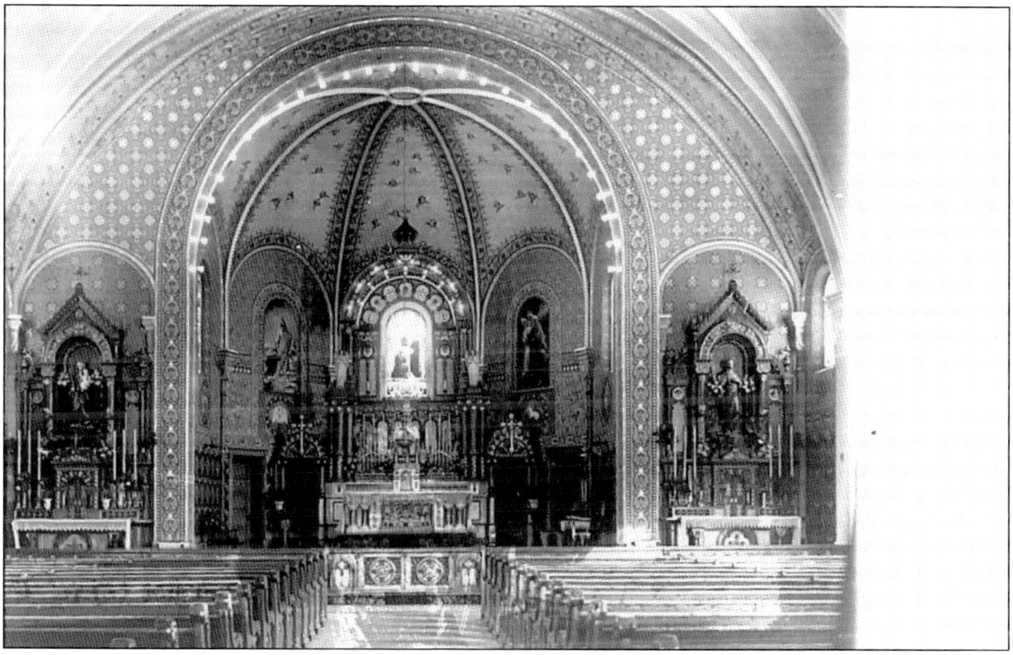

THE INTERIOR OF ST. PETER'S CHURCH. The altar of St. Peter's is pictured in 1910.

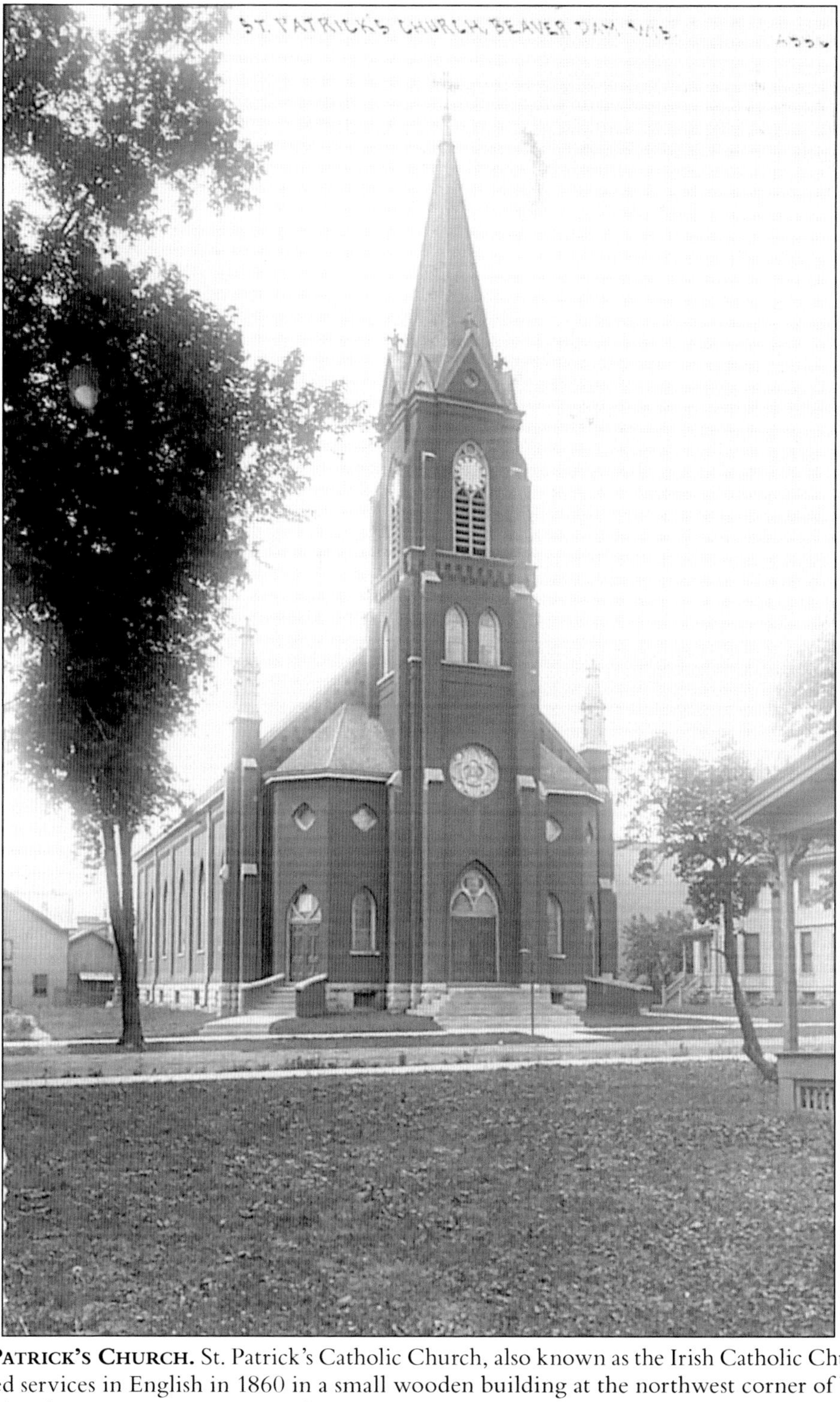

ST. PATRICK'S CHURCH. St. Patrick's Catholic Church, also known as the Irish Catholic Church, started services in English in 1860 in a small wooden building at the northwest corner of West Third and West Streets. In 1868, the congregation purchased the property at the corner of West Maple Avenue and North Center Street and erected its first church.

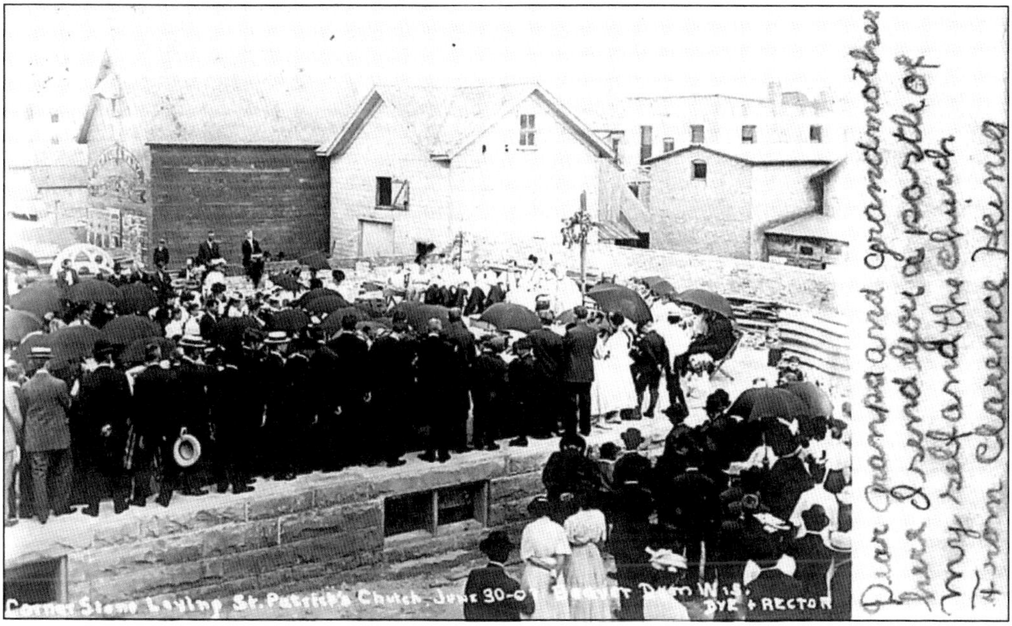

THE DEDICATION OF ST. PATRICK'S CHURCH. In 1907, St. Patrick's commenced the construction of the brick church pictured on the opposite page. The dedication of the new church took place on June 30, 1907.

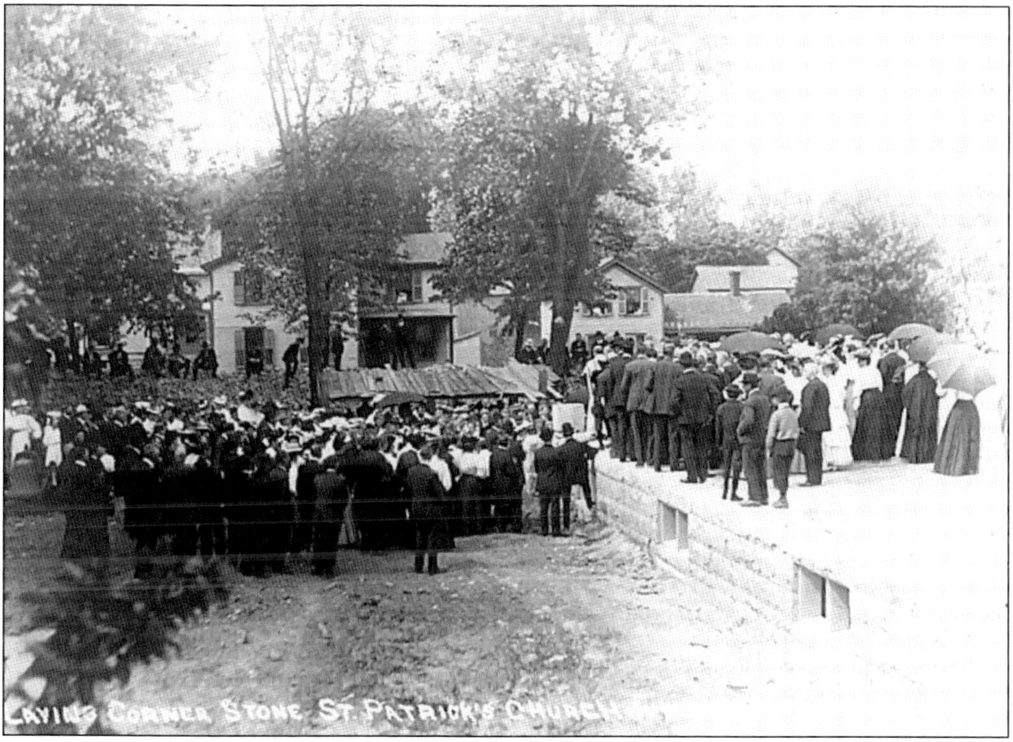

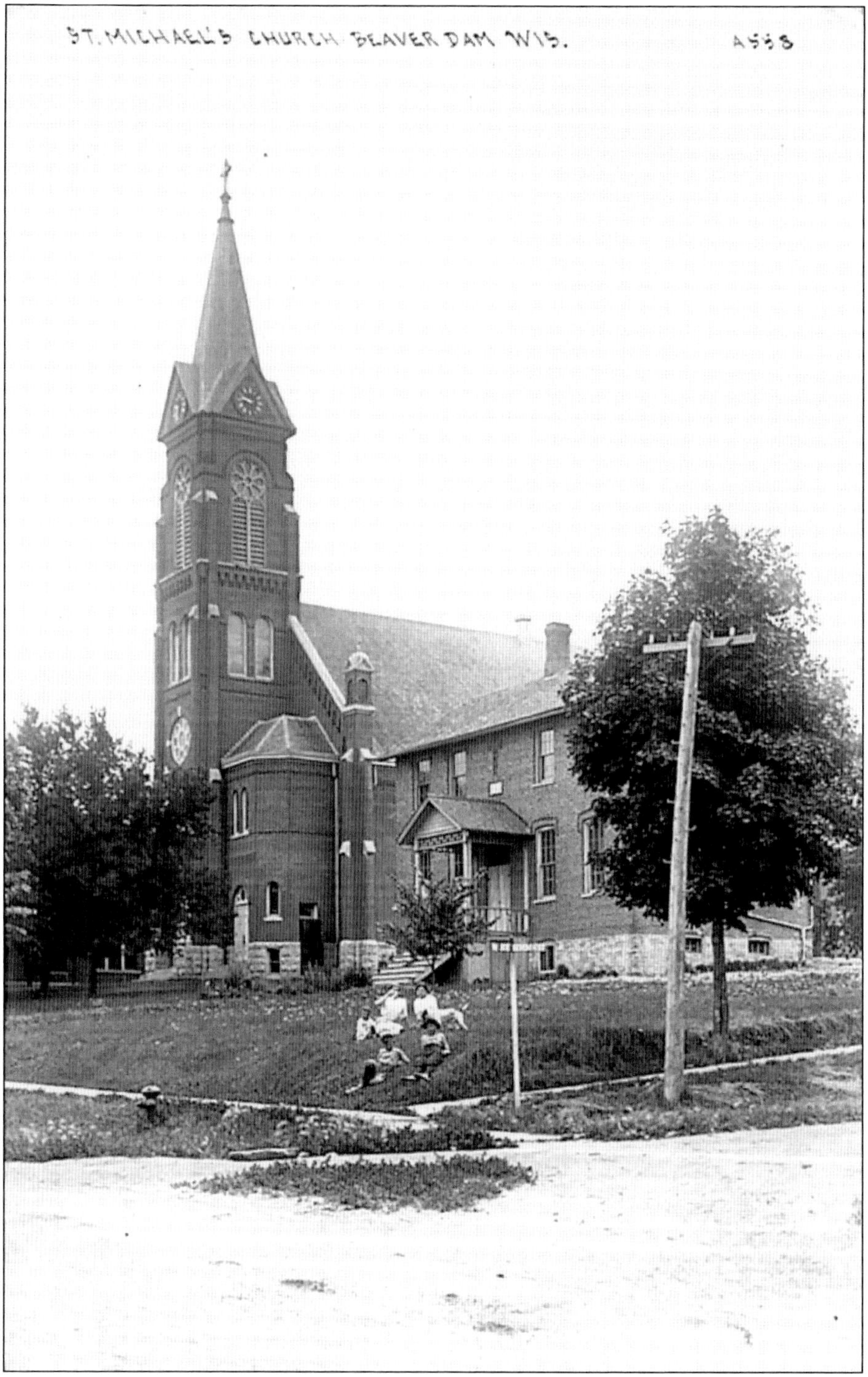

St. Michael's Catholic Church. St. Michael's Parish, also known as the Polish Catholic Church, took roots in the city in 1874. A small church was placed on the highest hill on Madison Street. In 1885, the congregation added a "little red school on the hill." In 1904, due to an increasing membership, the small church was replaced by the brick church shown above.

Nine

Dear Old Golden Rule Days

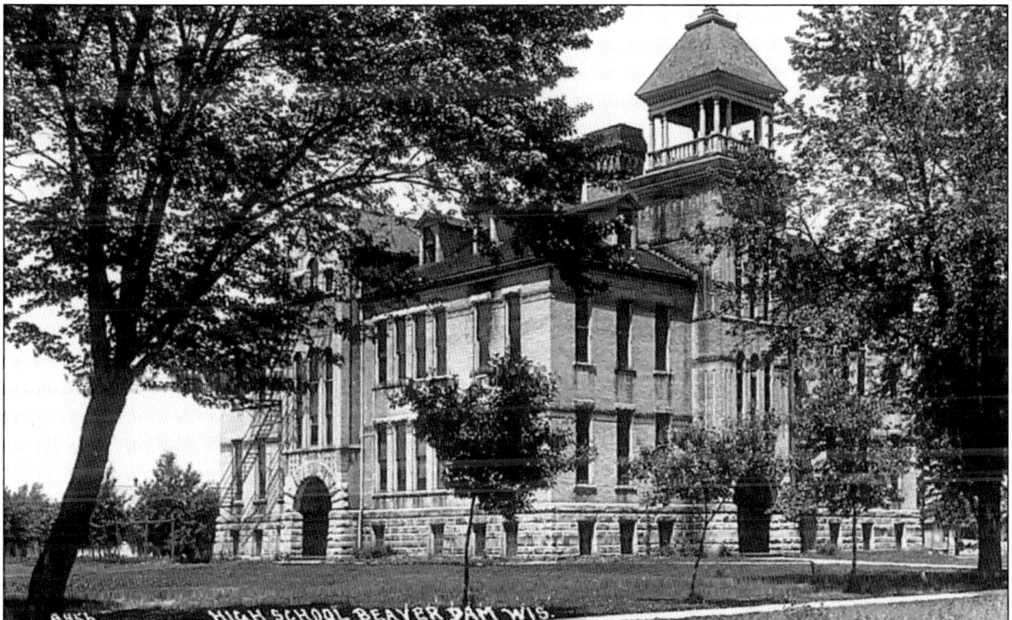

Beaver Dam High School. This school replaced the Union School as Beaver Dam's high school in 1891. Situated on the current site of Wilson School, it remained the high school until 1923. Subsequently, the building served as a grade school. It was razed in 1964.

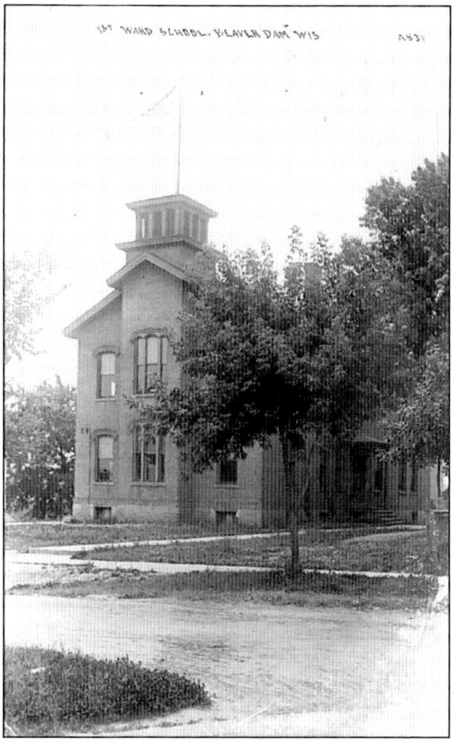

THE FIRST WARD SCHOOL. The First Ward School was situated at the corner of Mill and Madison Streets. The school was built in 1870, and an addition was built in 1876. As the ward designations were dropped, the First Ward School was renamed the Madison School.

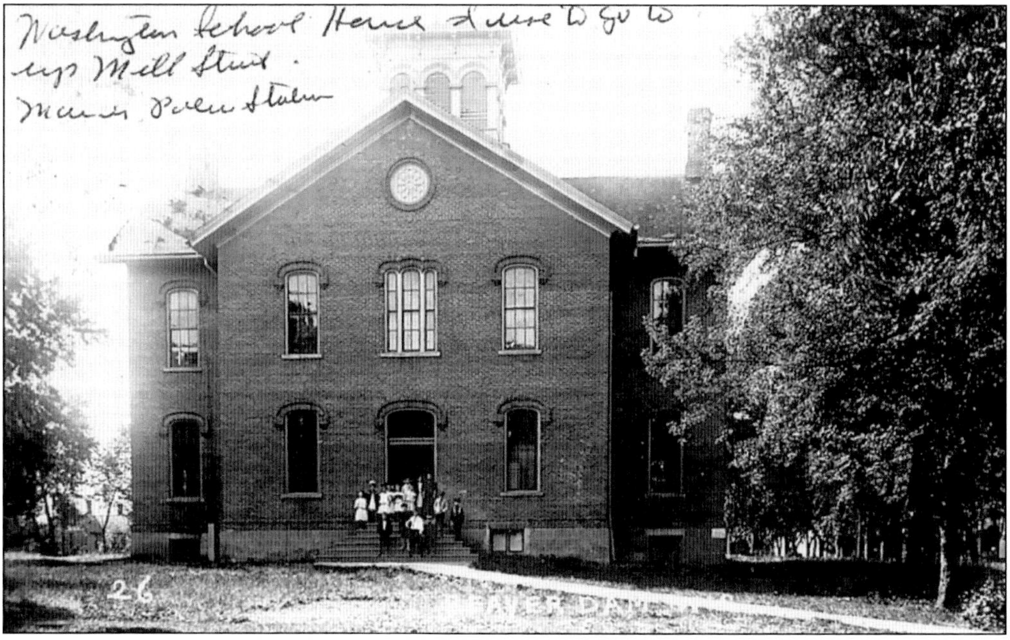

THE SECOND WARD SCHOOL. The Second Ward School, located at the corner of Lincoln and Washington Streets, was erected in 1871 as the Union School, which housed the high school and a grade school. Upon the construction of the new high school in 1891, the Second Ward School became Washington School and served solely as a grade school. The building was torn down in the early 1950s and replaced by the current city hall.

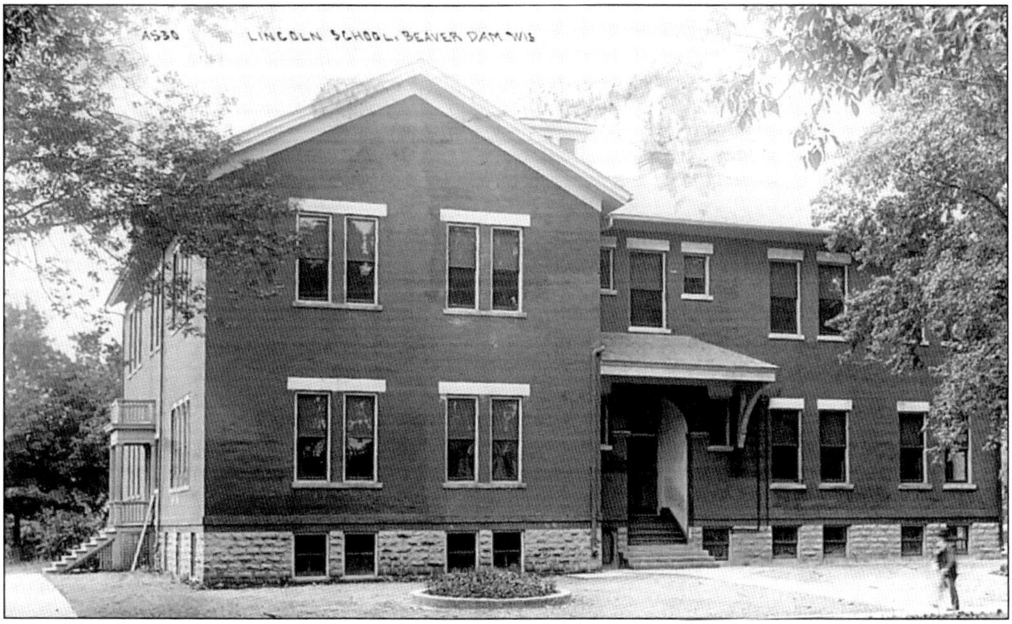

THE THIRD WARD SCHOOL. In 1911, the Third Ward School was erected, replacing a two-room brick school dating from 1867. In 1964, Lincoln School, as it was then called, met the wrecking ball and was replaced by Lincoln Park.

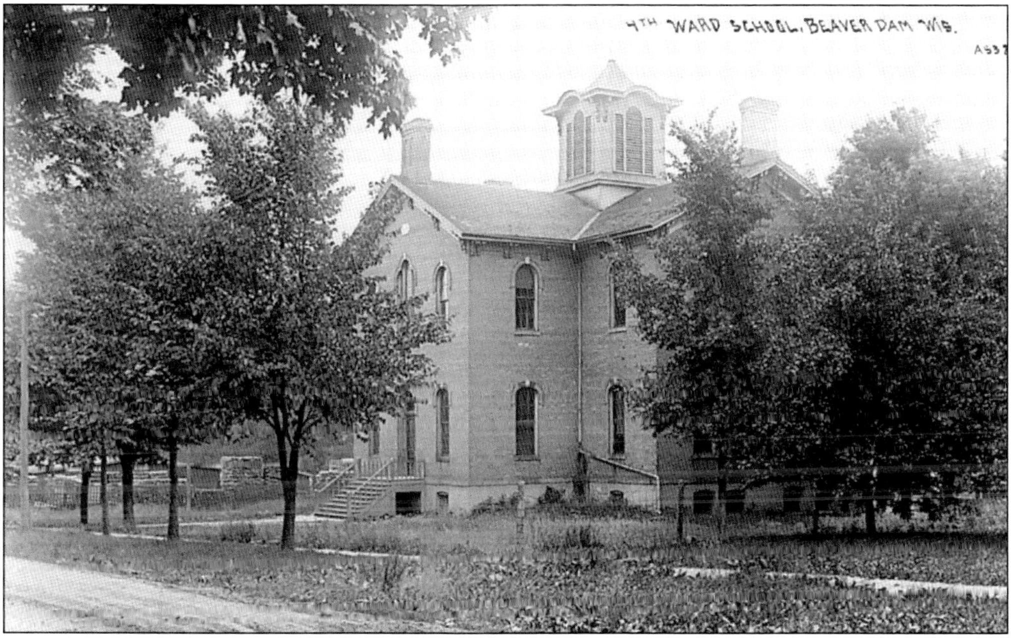

THE FOURTH WARD SCHOOL. The Fourth Ward School was built in 1867 at a site on Mackie Street between Spring and Center Streets. Upon the building of the new high school in 1923, the students of the Fourth Ward School were relocated to the Wilson School. The Fourth Ward School was thereafter used as the district's vocational and technical school.

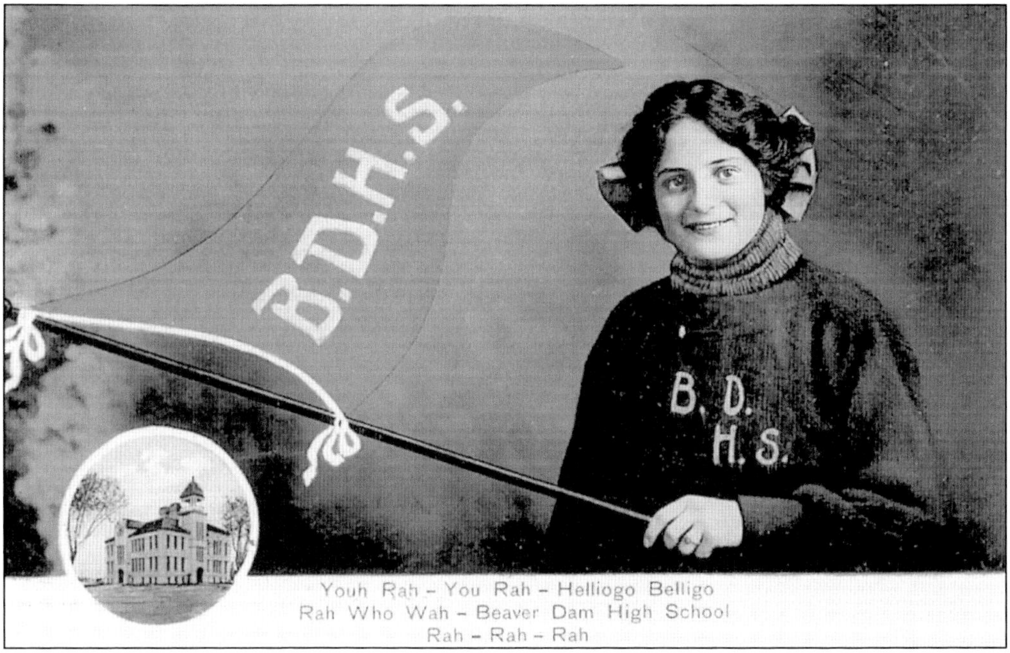

A CHEERLEADER POSTCARD. This generic postcard was used to promote many a high school—in this case, Beaver Dam High School. The German postcard maker could accommodate any school by using the high school initials, printing the pennant in the school color, and inserting a school photograph. A Wayland postcard was also available in this style.

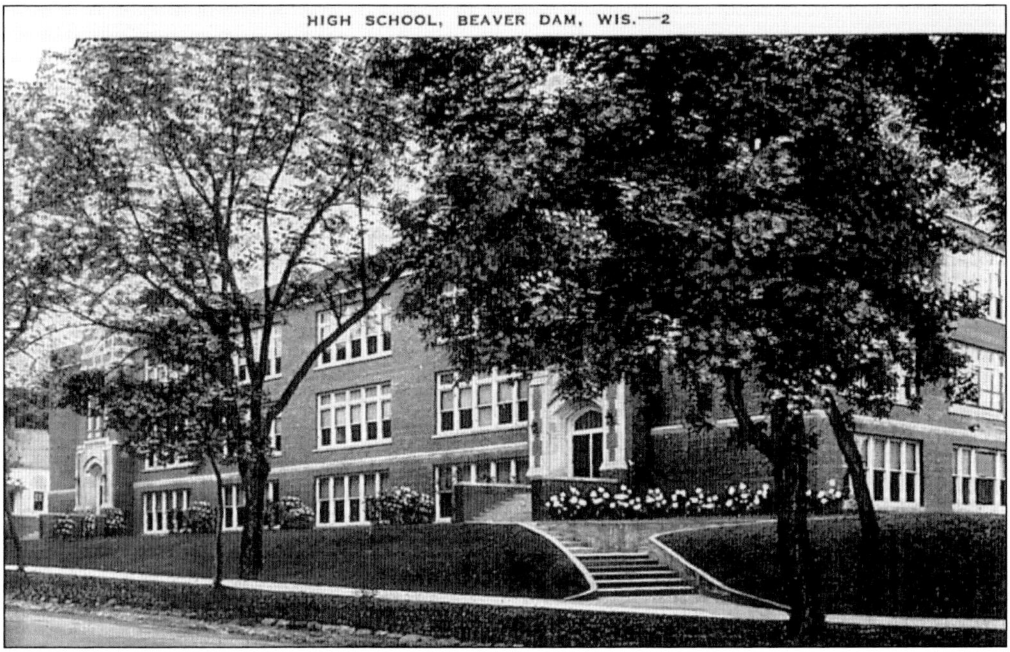

BEAVER DAM HIGH SCHOOL. In 1923, a new Beaver Dam High School opened on Fourth Street. It became the city's junior high school (later known as the middle school) in 1957 upon the opening of the current Beaver Dam High School on Gould Street.

A WAYLAND PENNANT POSTCARD. Like the one on the opposite page, this postcard design was also capable of accommodating any school. A similar postcard was produced to promote Beaver Dam High School.

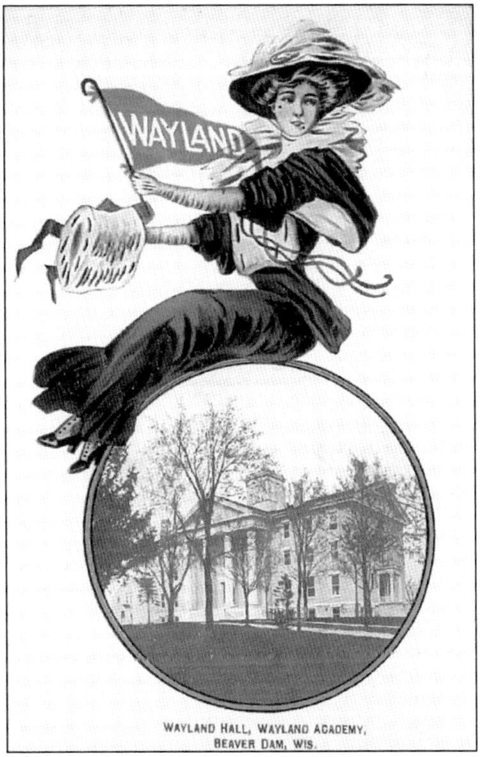

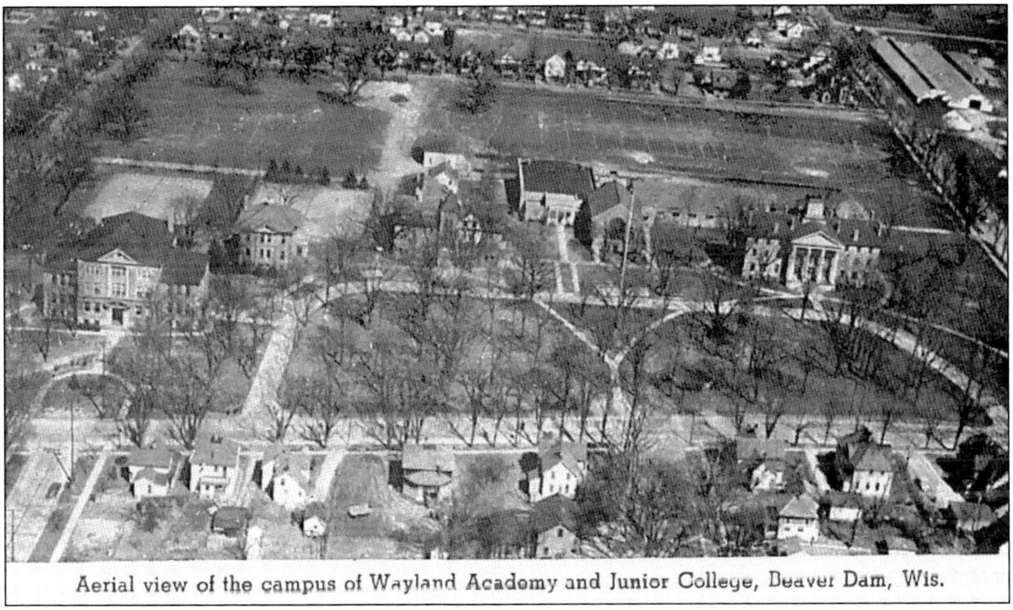

AN AERIAL VIEW OF WAYLAND ACADEMY AND JUNIOR COLLEGE. This school, established in 1855 as a Baptist boys' university, has experienced many changes over its 150-year history. It has been a college, a coed academy, and a junior college; it is now one of the finest preparatory schools in the nation. The aerial view afforded by this postcard shows the core of campus buildings. Shown around 1940 are, from left to right, Linfield Hall, Roundy Hall, Warren Cottage, Lindsay Gymnasium, and Wayland Hall.

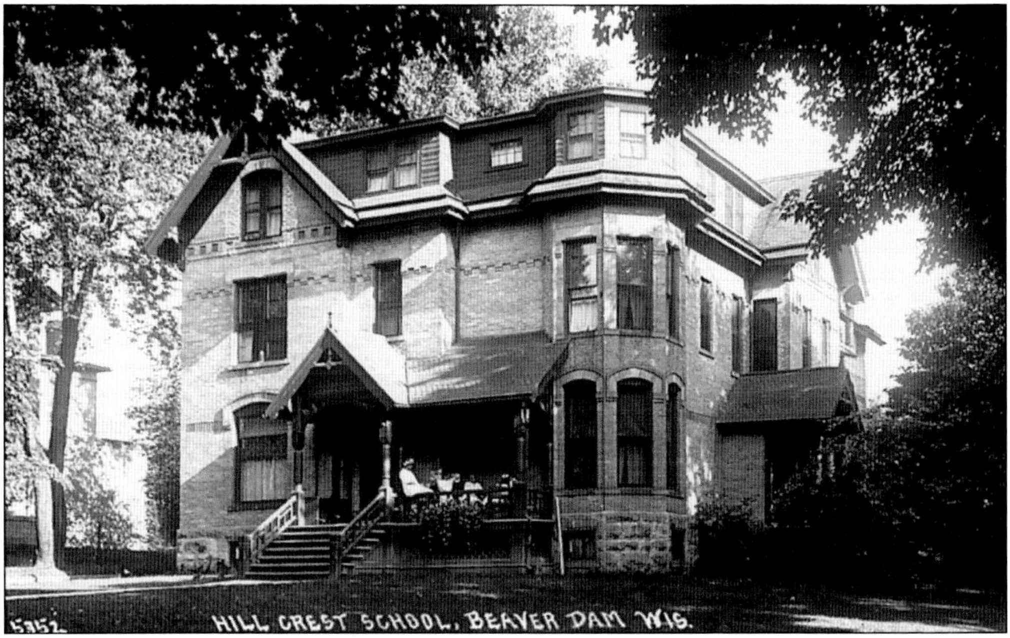

THE HILLCREST SCHOOL FOR GIRLS. Founded in 1910, the Hillcrest School for Girls was the dream of Sarah Davison. The school had many homes, but it finally settled in a complex of luxurious homes at the corner of Lincoln and Park Avenues. Pictured above is the first building, better known as the George Chandler home. The second building, pictured below, was the George Congdon home. Hillcrest was a private elementary school catering to girls from the ages of 6 to 14 years. The school did not survive the death of its creator in 1944, and the last graduation ceremony was held on June 6, 1947.

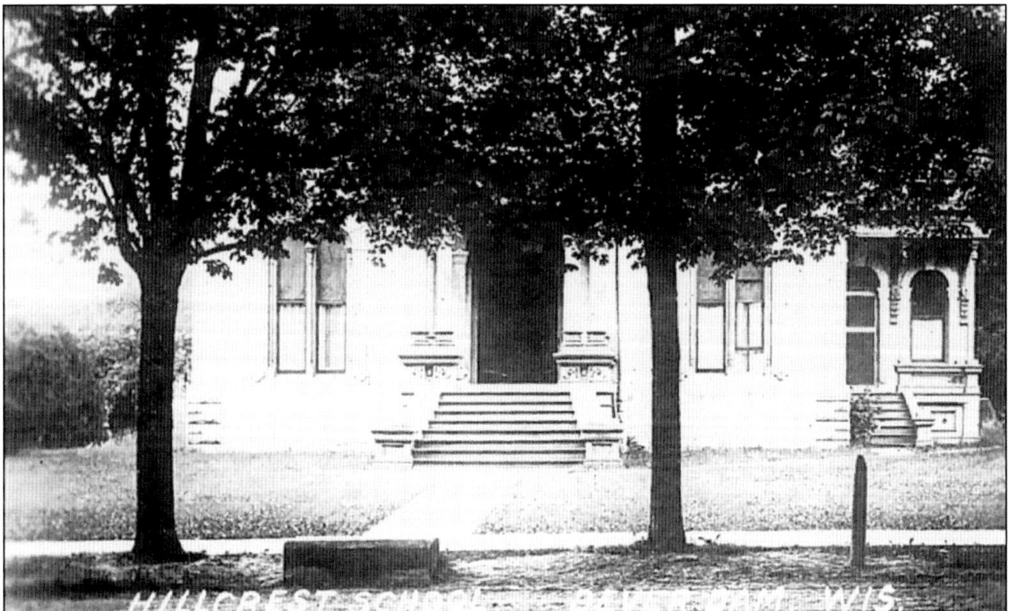

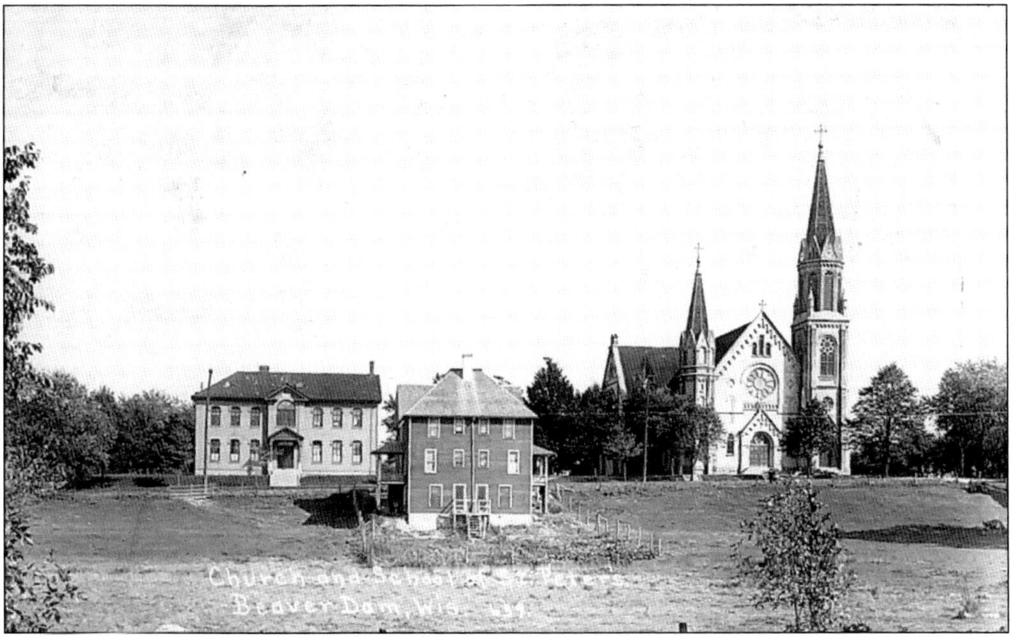

THE OLD ST. PETER'S CATHOLIC GRADE SCHOOL. St. Peter's School was founded in 1862. The school building shown above was erected in the 1870s.

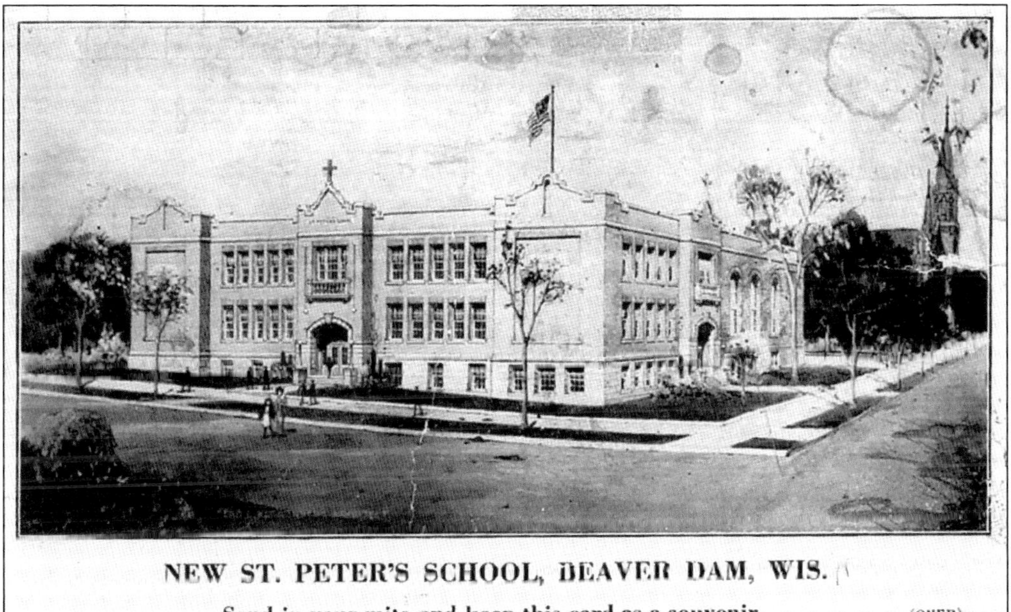

THE NEW ST. PETER'S SCHOOL. In 1922, this postcard announced the construction of a new St. Peter's School and solicited for donations. The school was substantially enhanced in 1954 with the addition of eight more classrooms, a gym, a cafeteria, a kitchen, a library, and a faculty room.

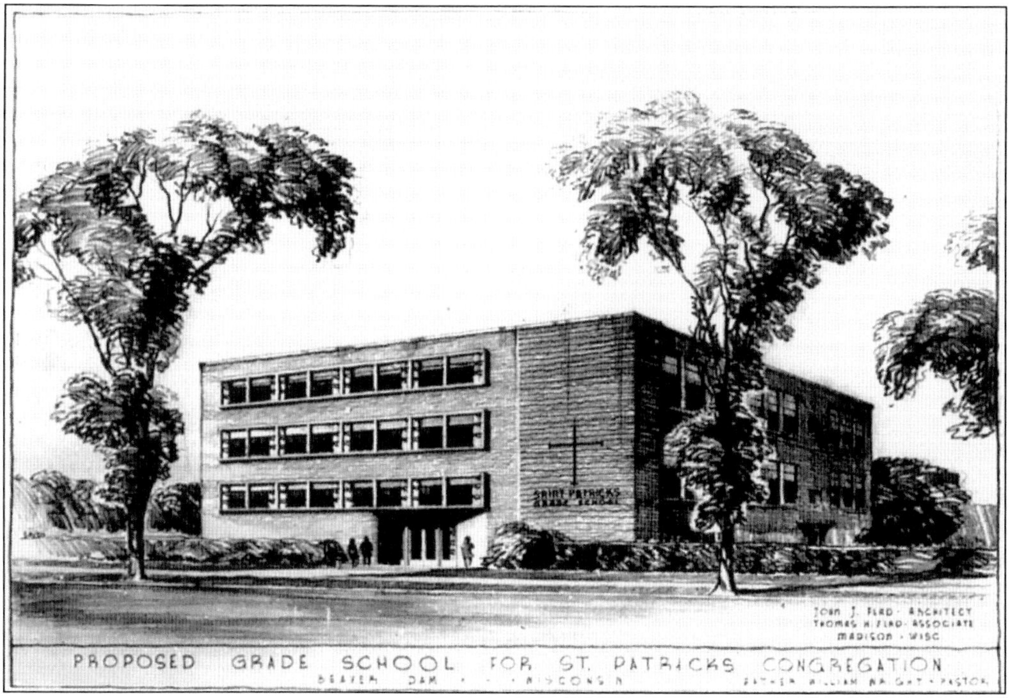

St. Patrick's Catholic Grade School. This postcard, which is similar to the one seen on the previous page, was created nearly 30 years later to announce the opening of St. Patrick's Grade School in 1950.

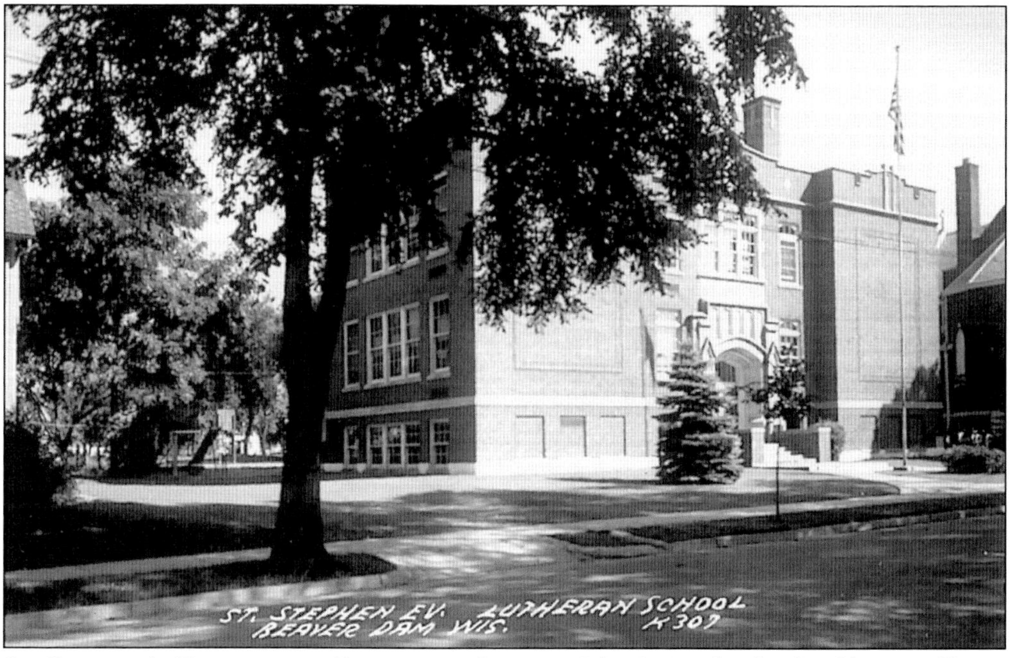

St. Stephen's Evangelical Lutheran School. The St. Stephen's school began offering instruction in its old church in 1896. Two school building projects followed, one in 1913 and one in 1929. The 1929 school is pictured above.

Ten

Do You Believe in Magic?

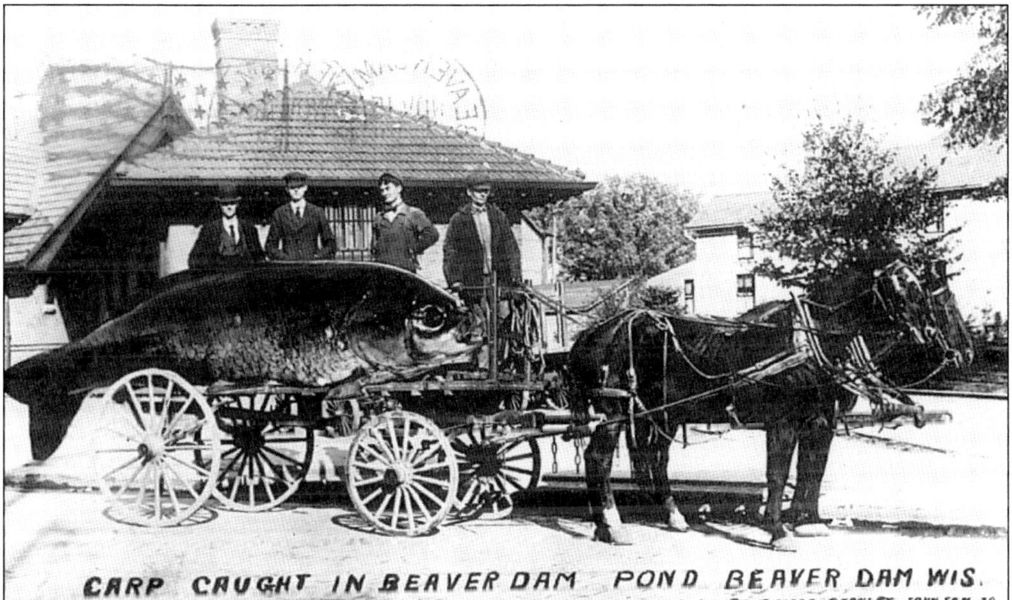

Carp at the Depot. A popular style of novelty postcards in the early 20th century involved trick photography. The postcard maker chose a local scene as the backdrop and superimposed an exaggerated object onto it. One of the country's finest proponents of this postcard lived just up the road from Beaver Dam. Alfred Stanley Johnson Jr. of Waupun created these types of postcards for towns and cities throughout the nation. This particular postcard uses the backdrop of the downtown depot to extol the size of fish caught in Beaver Dam Lake.

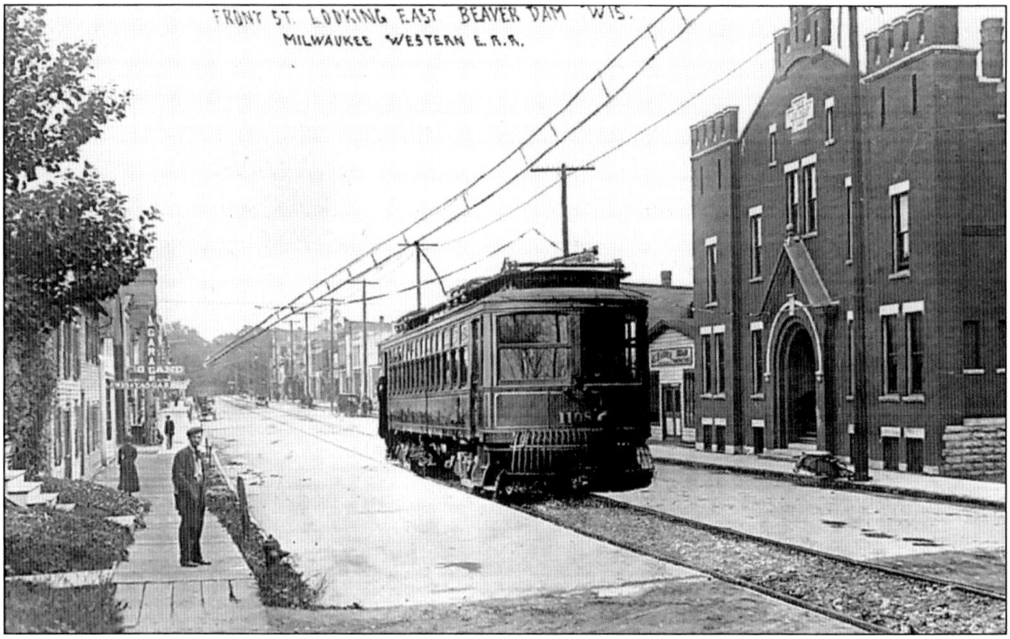

THE FRONT STREET TROLLEY, DOWNTOWN. Another common subject of the trick-photography postcard was the electric trolley car. This image of the area in front of the armory gives the viewer a mistaken impression about the size and advancement of the city's public transportation.

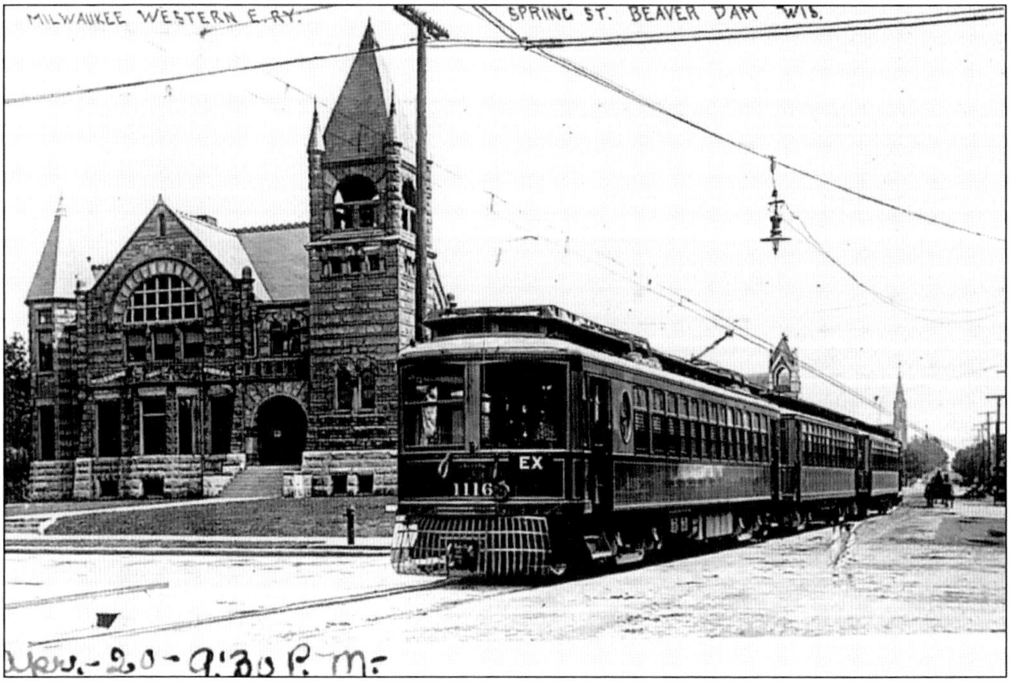

THE SPRING STREET TROLLEY. On this trick-photography postcard, a second trolley line has been created in front of the Williams Free Library.

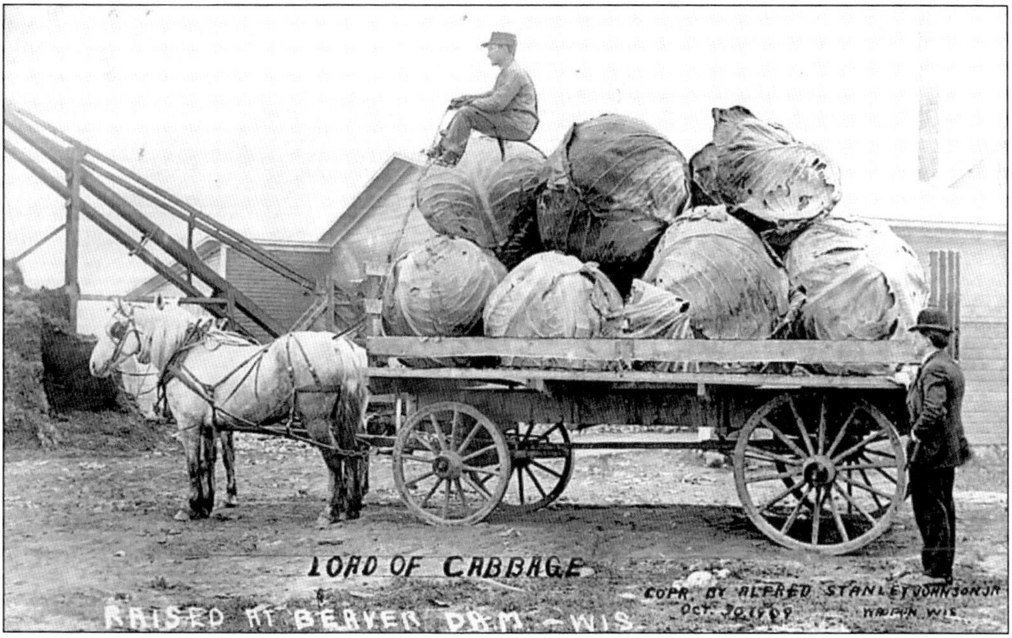

CABBAGE FARMING. Embellishing the size of crops was also common on trick-photography cards.

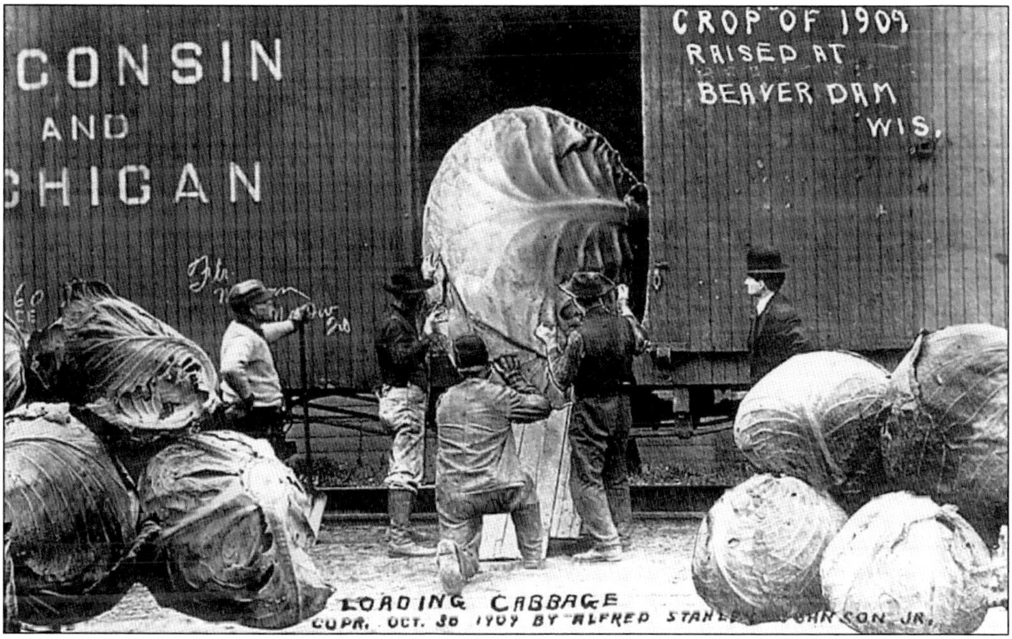

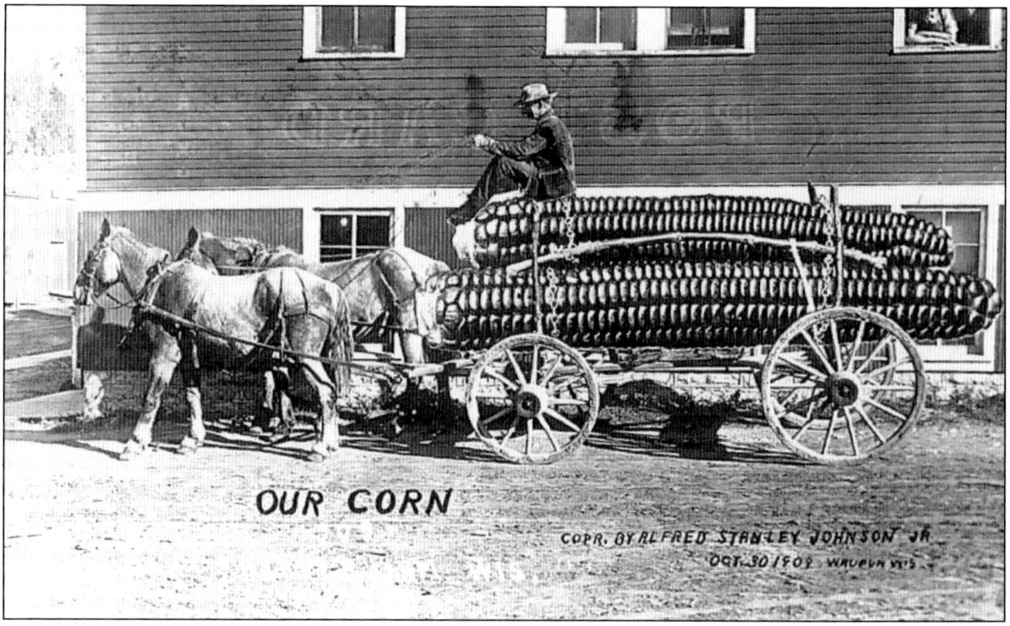

CORN CROPS.

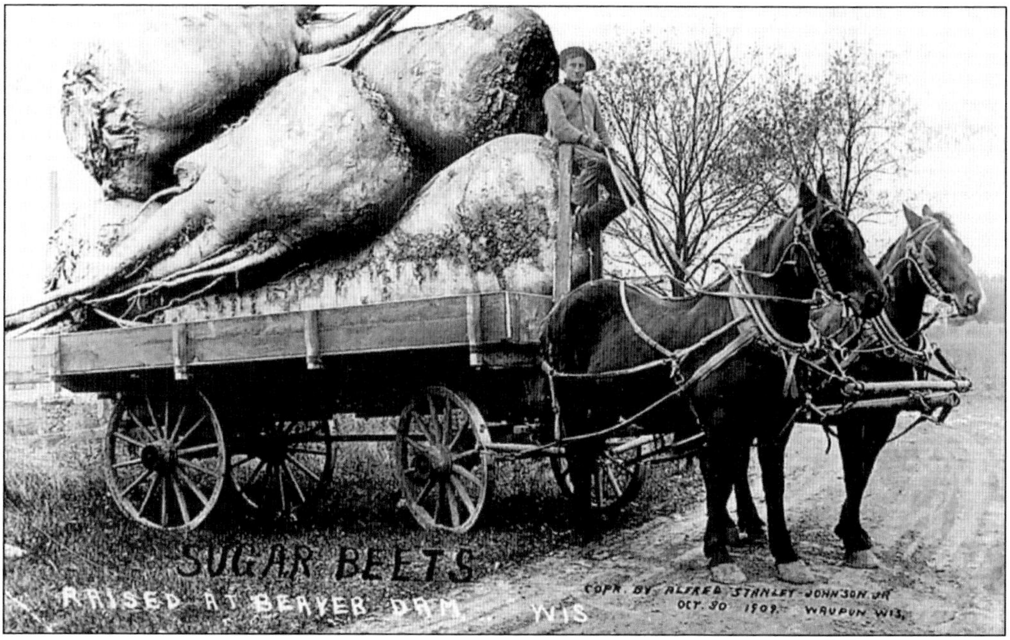

HARVESTING THE SUGAR BEET.

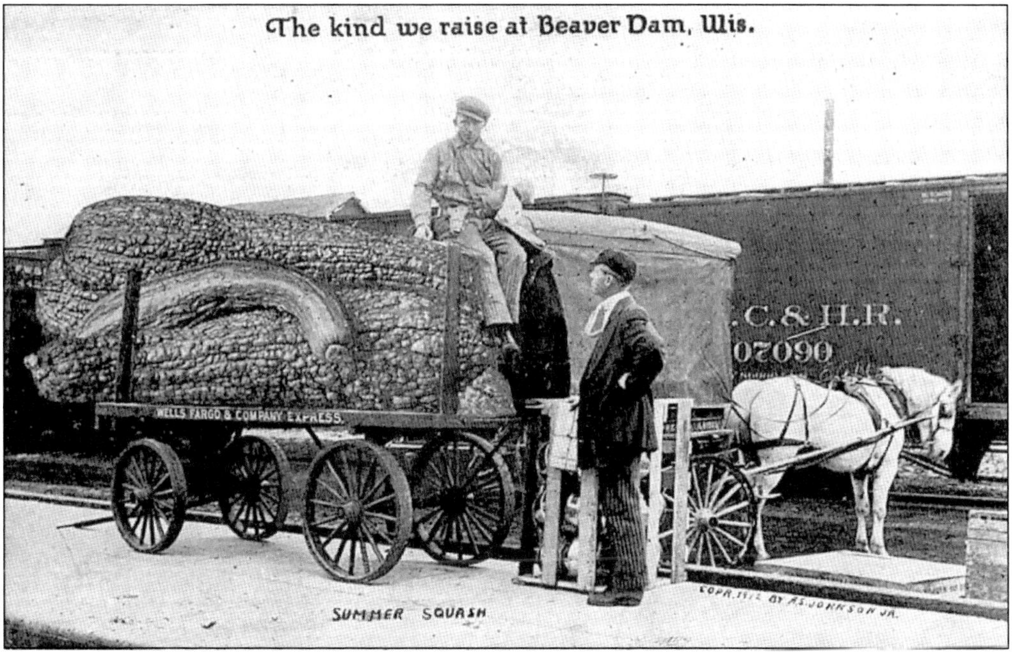

SUMMER SQUASH AT THE NORTH SPRING STREET DEPOT.

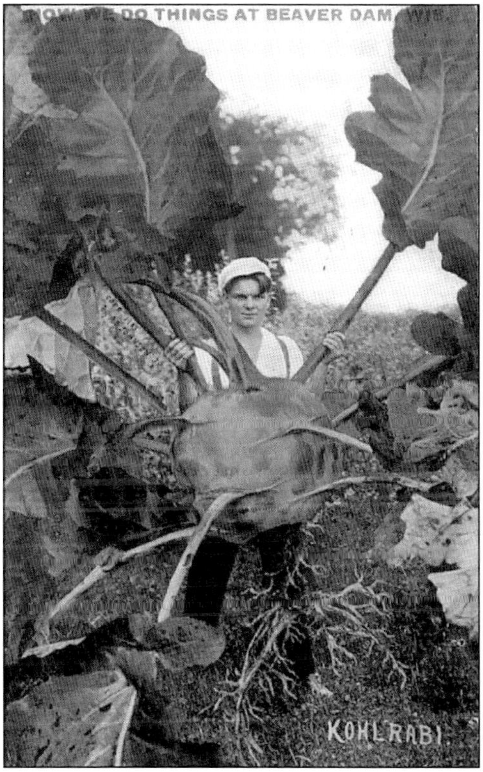

KOHLRABI.

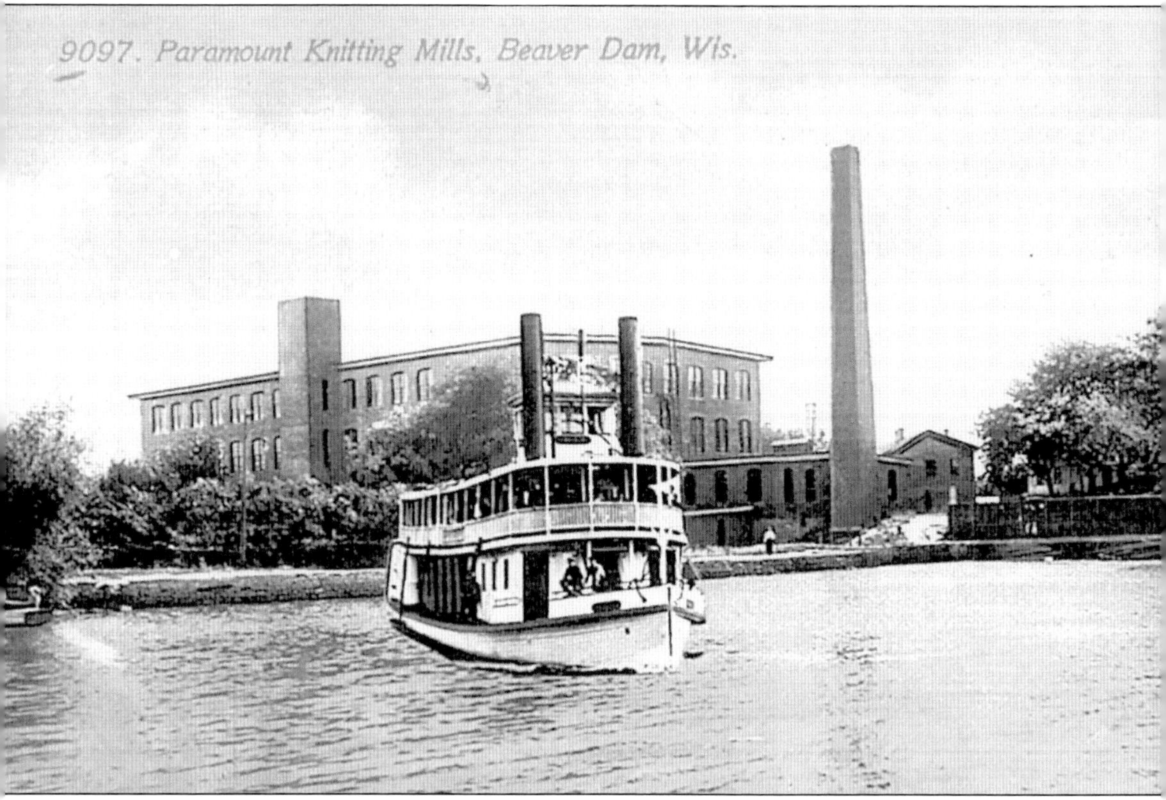

A RIVERBOAT ON BEAVER DAM LAKE. Beaver Dam Lake was frequented by many tour boats in the 1890s. But with a capacity of 25 to 30 patrons, they could not compare to the illusion of the Mississippi-style steamboat created on this postcard.

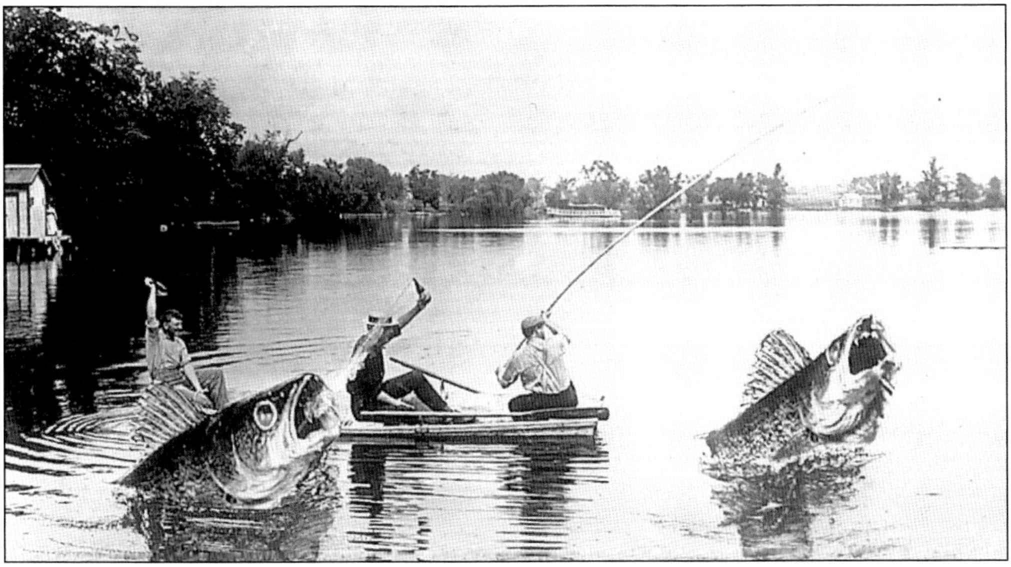

FISHING AT BEAVER DAM LAKE. Beaver Dam Lake has always been a favorite tourist spot to drop a line, but no one went home with fish this size.

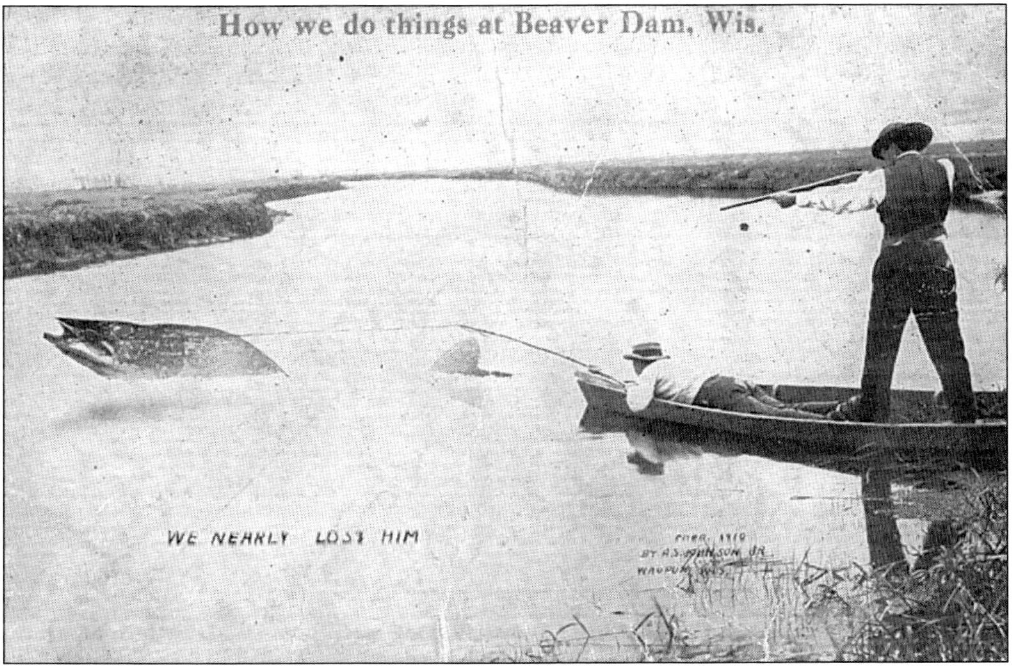

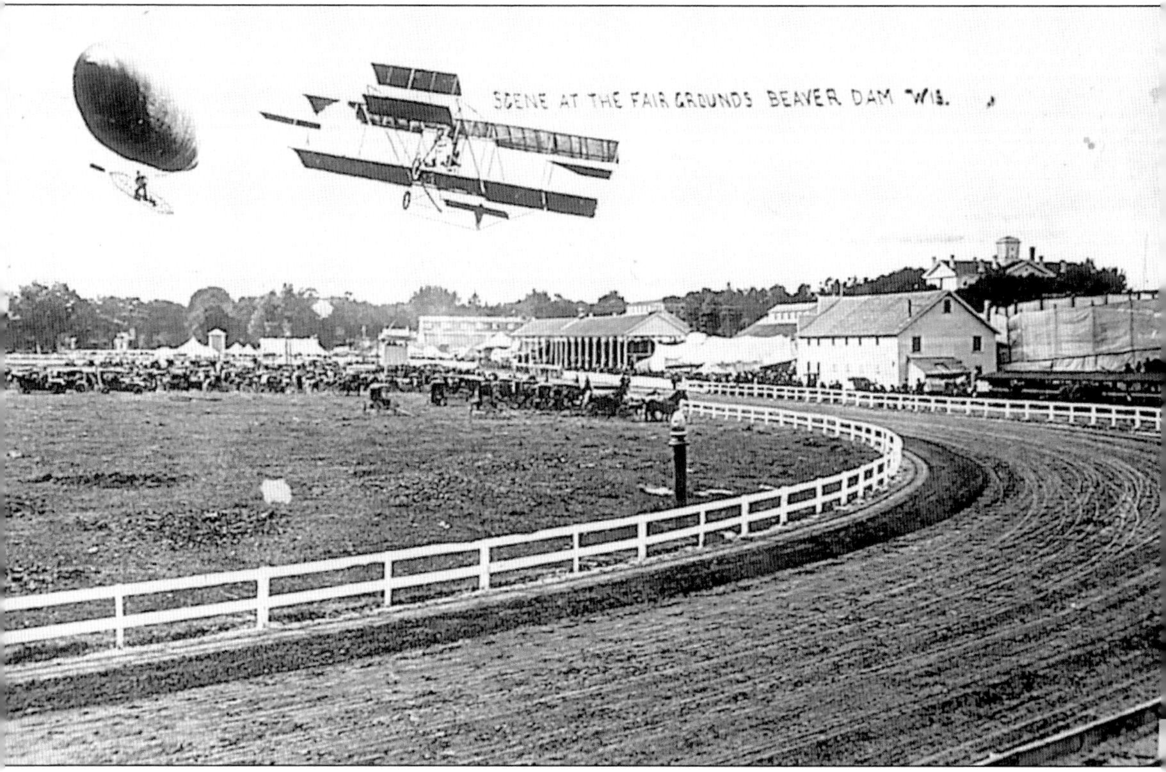

EVERYONE'S COMING TO THE FAIR. The first airplane to arrive in Beaver Dam landed at the Dodge County Fair. But it was not the plane or zeppelin shown in this postcard. The biplane shown here was superimposed on several postcards from Beaver Dam. The postcard maker placed it soaring above the lake, downtown buildings, and rows and rows of cornfields.

Bibliography

Beaver Dam Historical Committee. *Centennial History Beaver Dam, Wisconsin.* Beaver Dam Centennial Inc., 1941.

Beaver Dam History Book Committee. *Sesquicentennial History Beaver Dam, Wisconsin: 1841–1991.* Beaver Dam Sesquicentennial, 1991.

History of Dodge County. Chicago: Western Historical Company, 1880.

Hubbell, Horner Bishop. *Dodge County Wisconsin Past and Present.* Chicago, Illinois: S. J. Clark Publishing Company, 1913.

Neis, Stefano. "A Brief History of Postcard Types." Home page. May 13, 2001, June 2, 2005 http://www.geocities.com/Heartland/Meadows/2487/pchistory.htm.

People. . . Places. . . Things. . . Of Yesteryear in Beaver Dam. Beaver Dam, Wisconsin, 1961.